Lecture Notes in Mathematics

A collection of informal reports and seminars
Edited by A. Dold, Heidelberg and B. Eckmann, Zürich

Series: Department of Mathematics, University of Maryland, College Park
Adviser: J. K. Goldhaber

210

Martin Eichler
Universität Basel, Basel/Schweiz

Projective Varieties and Modular Forms

Course Given at the University of Maryland, Spring 1970

Springer-Verlag
Berlin · Heidelberg · New York 1971

AMS Subject Classifications (1970): 18G10, 14M05, 10D20

ISBN 3-540-05519-3 Springer-Verlag Berlin · Heidelberg · New York
ISBN 0-387-05519-3 Springer-Verlag New York · Heidelberg · Berlin

Offsetdruck: Julius Beltz, Hemsbach/Bergstr.

CONTENTS

INTRODUCTION

W.L. Baily and *A. Borel* showed (Compactification of arithmetic quotients of bounded symmetric domains, Annals of Math. **84**(1966), p. 442-528) that the graded rings of automorphic forms are finitely generated, provided that rather general conditions are satisfied. Let $Y_0,\cdots,Y_m$ be a system of generators, then between the Y_i some polynomial equations $F_\nu(Y_0,\cdots,Y_m) = 0$ will hold, and all these equations are consequences of finitely many of them. Let us assume for the sake of simplicity (although it is wrong in general) that the Y_i have the same weights. Then the $F_\nu(Y_i)$ are homogeneous polynomials, and their zeros form a *projective variety* $\mathbf{G}^n$ in the projective space $\mathbf{P}^m$ with the coordinates Y_i, whose dimension is $n \leq m$. Projective varieties are the most common examples of algebraic varieties although it can be shown that there exist (rather exceptional) algebraic varieties which are not projective. The theorem of Baily and Borel opens the way to an algebraic geometric treatment of automorphic forms and functions.

A central theorem in Algebraic Geometry, i.e., in the theory of algebraic varieties, is the theorem of Riemann-Roch. In the case of dimension $n = 1$ there exist in principle two proofs: that of *Dedekind* and *Weber* (1882), and *A. Weil*'s proof of 1938. Dedekind and Weber use a projective model of the algebraic curve, and they obtain first a more general theorem on modules over a polynomial ring which they later apply to the special situation. The disadvantage of their method is that it does not make it clear from the beginning that their result is independent of the model, in other words, that it is an invariant of the underlying abstract algebraic variety $\mathbf{G}$. This invariance property has to be established later. Weil's proof works right from the beginning with invariant concepts. Its generalization to dimensions $n > 1$ uses sheaf theory. Under the most general assumptions we find if in *J.-P. Serre*'s paper "Faisceaux Algébriques Cohérents," Annals of Math. 61(1955), p. 197-278. The theorem of Riemann-Roch contains a number of invariants of the variety which are in general hard to determine; and unless the variety has no "singular points," the sheaf theoretical access seems today extremely difficult.

The varieties defined by modular forms in many variables have singular points, and so they represent serious obstacles for the application of sheaf theoretic methods. Fortunately the generalization of Dedekind and Weber's approach to a dimension $n > 1$ is not obstructed by singularities. In these lectures we shall give a self-contained introduction to the theory of projective varieties, following the direction set by Dedekind and Weber.

Chapter I develops the formal tools, the theory of graded modules over polynomial rings. Chapter II applies these tools to projective varieties. Chapter III is devoted to the determination of the terms in the theorem of Riemann-Roch for the varieties of Siegel and Hilbert modular forms, and to some implications of that. At the root of these varieties there lie arithmetical objects, namely the modular groups, therefore number theory must mix with algebraic geometry in order to yield definite results. Here we see a new encounter between different mathematical branches of which the history of modular functions has given striking examples. We shall end up by mentioning a number of open problems in the appendix.

For the understanding of Chapters I and II, no special knowledge is required. Chapter III can also be read without being familiar with the theory of modular forms, if one is prepared to accept without proof some statements on modular forms and quadratic forms.

This chapter has been materially changed after the lectures

because of an error in the proof of what now appears as Hypothesis in §16. Shortly later we found a correct proof. It will appear in the ACTA ARITHMETICA under the title: Über die graduierten Ringe der Modulformen. This proof involves quite a few deep tools in number theory and, in a first reading, it should be skipped in any case. So it is a happy coincidence that it does not appear in these Lecture Notes. Already now it will become clear that our approach to Algebraic Geometry is in some way superior to others, as it allows us to deal with singular varieties directly, without desingularization, which is a tedious task in practical cases.

CHAPTER I

GRADED MODULES

§1. *Some basic concepts*

In §1-§3 h is an arbitrary ring with unit element 1 (later h will be specialized). An h-module M is a commutative additive group with the following property: for elements $h_1, h_2, \cdots \in h$ and $M_1, M_2, \cdots \in M$ a product is defined satisfying

1) $1 \cdot M = M$, 3) $h(M_1+M_2) = hM_1 + hM_2$,

2) $(h_1+h_2)M = h_1M + h_2M$ 4) $(h_1h_2)M = h_1(h_2M)$.

An element $M \neq 0 \in M$ is called a *torsion element*, if an element $h \neq 0 \in h$ exists such that $hM = 0$. A module without torsion elements is called *torsion-free*. If h is commutative and without divisors of zero, the torsion elements together with 0 form the *torsion submodule* M_0. (In this case the quotient module

$$M_1 = M/M_0$$

is also a h-module, the *torsion-free kernel of* M.)

Let N be another h-module. A homomorphic map $\mu : M \longrightarrow N$ of M into N is called *h-linear* if for all $h_1, h_2 \in h$ and $M_1, M_2 \in M$:

$$\mu(h_1M_1+h_2M_2) = h_1\mu M_1 + h_2\mu M_2.$$

We shall not speak of maps which are not h-linear and therefore always omit this word. μ is *surjective* if all elements of N are images. μ is *injective* if it maps M isomorphically onto a submodule of N. μ is *bijective* or an isomorphism if it is both surjective and injective.

All these maps form also an h-module which is written $\mathrm{Hom}_h(M,N)$. Addition and multiplication by elements $h \in h$ is defined by

$$(\mu_1+\mu_2)M = \mu_1 M + \mu_2 M, \qquad (\mu h)M = \mu(hM).$$

The difference between M and $\mathrm{Hom}_h(M,N)$ is that the former is a *left module* and the latter a *right module*.

The following theorem is almost obvious:

Theorem 1. $\mathrm{Hom}_h(M_1 \oplus M_2, N) = \mathrm{Hom}_h(M_1,N) \oplus \mathrm{Hom}_h(M_2,N)$.

If h is commutative, which will be assumed from now on, we need not distinguish between left and right modules. In particular we may write $h\mu$ instead of μh.

h is a *graded ring* if it is the direct sum of additive subgroups h^i $(i = 0, \pm 1, \pm 2, \cdots)$ which may be partly empty, for which $h^i h^j \subseteq h^{i+j}$. Similarly an h-module M is *graded* if

$$M = \oplus M^i, \qquad h^i M^j \subseteq M^{i+j}.$$

The elements of M^i (or h^i) are the *homogeneous elements of degree* i of M (or h).

Proposition 1. *If h, M, N are graded and h is commutative, then $\mathrm{Hom}_h(M,N)$ is also graded. The homogeneous elements $\mu^i \in \mathrm{Hom}_h(M,N)$ map the homogeneous elements $M^j \in M$ on homogeneous elements $\mu^i M^j \in N$ of degree $i+j$.*

Proof. A $\mu \in \mathrm{Hom}_h(M,N)$ is uniquely determined by its action on the $M^i \in M^i$. Let

$$\mu M^j = \sum_i N^{i+j} \tag{1}$$

and define μ^i by

$$\mu^i M^j = N^{i+j}. \tag{2}$$

Then the sum

$$\mu = \sum_i \mu^i \tag{3}$$

satisfies (1) which proves (2). The result can be written

$$\mathrm{Hom}_h(M,N) = \bigoplus \mathrm{Hom}_h(M,N)^i , \tag{4}$$

(4) and (2) together are the statements made. But it remains to be proved that the μ^i defined by (1) and (2) are h-linear.

Let $h_1^k, h_2^k \in h^k$, $M_1^j, M_2^j \in M^j$. Then $h_1^k M_1^j + h_2^k M_2^j \in M^{j+k}$ and

$$\sum_i \mu^i (h_1^k M_1^j + h_2^k M_2^j) = h_1^k \sum_i \mu^i M_1^j + h_2^k \sum_i \mu^i M_2^j .$$

All summands on the left and right are homogeneous elements of degrees $i+j+k$ of N_1 and because of the uniqueness of the decomposition of N:

$$\mu^i (h_1^k M_1^j + h_2^k M_2^j) = h_1^k \mu^i M_1^j + h_2^k \mu^i M_2^j ,$$

QED.

In the case of graded rings and modules we are only interested in the homogeneous elements. Therefore we restrict addition to homogeneous elements of equal degrees. Under this convention we can always omit the word "homogeneous."

The following theorem is an easy adaptation of a well-known theorem to the graded case.

Theorem 2. Let h, M, N be graded and h commutative, and $\sigma : M \longrightarrow N$ *a surjective homomorphism of degree s. Then the kernel K of σ is a graded submodule of M, and there exists a homomorphism* $\rho : M \longrightarrow M/K$ *of degree 0 and an isomorphism* $\varepsilon : M/K \longrightarrow N$ *of degree s such that* $\sigma = \varepsilon\rho$.

The modules occurring in this theorem are put together in the diagram

(5) $$0 \xrightarrow[\mu]{} N \xrightarrow[\iota]{} M \xrightarrow[\sigma]{} Q \xrightarrow[\nu]{} 0,$$

where the trivial maps μ of the zero-module into N and ν of Q onto the zero-module are added. The fact that N is a submodule of M can be expressed as follows: there exists an injection ι of N into M. The product $\iota\mu$ acting on 0 is of course $\iota\mu = 0$. The fact that ι is an injection is equivalent to the statement that *only* the image $\mu 0$ of 0 is mapped on 0 in M. So we have

$$\text{im } \mu = \ker \iota.$$

Similarly we have $\sigma\iota N = 0$ or briefly $\sigma\iota = 0$, and by Theorem 2:

$$\text{im } \iota = \ker \sigma.$$

Lastly we have $\nu\sigma = 0$ which is trivial and by Theorem 2:

$$\text{im } \sigma = \ker \nu.$$

So (5) is an example of an *exact sequence*

$$\cdots \longrightarrow M_{i-1} \xrightarrow[\alpha_{i-1}]{} M_i \xrightarrow[\alpha_i]{} M_{i+1} \xrightarrow[\alpha_{i+1}]{} \cdots$$

of h-modules M_i where

$$\text{im } \alpha_{i-1} = \ker \alpha_i, \quad \text{in particular} \quad \alpha_i \alpha_{i-1} = 0.$$

If only $\alpha_i \alpha_{i-1} = 0$, we shall speak of a *complex* of modules.

Proposition 2. *If* $0 \xrightarrow[\alpha_0]{} M_1 \xrightarrow[\alpha_1]{} M_2 \longrightarrow \cdots$ *is an exact sequence, a map* $\alpha_2' : M_2/\text{im } \alpha_1 \longrightarrow M_3$ *can be defined such that*

$$0 \xrightarrow[\alpha_0']{} M_2/\text{im } \alpha_1 \xrightarrow[\alpha_2']{} M_3 \xrightarrow[\alpha_3]{} M_4 \longrightarrow \cdots$$

is also an exact sequence.

Proof. The exactness of the given sequence at M_1 means that $\text{im } \alpha_1$ is a submodule of M_2 which is isomorphic with M_1. The exactness at M_2 states that this submodule and no larger one is mapped by α_2 into the zero of M_3. So α_2 defines a mapping $\alpha_2' : M_2/\text{im}\ \alpha \longrightarrow M_3$ in a natural way, and $\text{im}\ \alpha_2' = \ker \alpha_3$ follows immediately from the exactness of the given sequence at M_3.

Later we shall use the following application of Proposition 2:

Proposition 3. *Let*

$$0 \xrightarrow[\alpha_0]{} M_1 \xrightarrow[\alpha_1]{} \cdots \longrightarrow M_n \xrightarrow[\alpha_n]{} 0 \tag{6}$$

be a finite exact sequence of h-modules beginning and ending with the zero-module. Assume h and the M_i *to be graded, the maps* α_i *to be of degree* 0*, the submodule* h^0 *of elements of degree* 0 *in h to be a field, and the submodules* M_i^λ *of degrees* λ *of the* M_i *to have finite dimensions* $L(\lambda, M_i)$ *over* h^0*. Then*

$$\sum_{i=1}^{n} (-1)^i L(\lambda, M_i) = 0. \tag{7}$$

Proof by induction on n. For $n = 2$ the exactness of (6) means that α_1 is an isomorphism of degree 0, and (7) is clear. For $n = 3$, (6) states the isomorphism

$$M_3 \cong M_2/\text{im}\ \alpha_1$$

which is of degree 0. In this case,

$$\begin{aligned} L(\lambda, M_3) &= L(\lambda, M_2) - L(\lambda, \text{im}\ \alpha_1) \\ &= L(\lambda, M_2) - L(\lambda, M_1) \end{aligned} \tag{8}$$

which is (7). For larger n we apply Proposition 2 and (8) which leads to the induction

$$\begin{aligned} &L(\lambda, M_1) - L(\lambda, M_2) + L(\lambda, M_3) - + \cdots \\ &\qquad = -L(\lambda, M_2/\text{im}\ \alpha_1) + L(\lambda, M_3) - + \cdots = 0. \end{aligned}$$

§2. *Free resolutions*

In §2 no special assumptions on the ring h and the h-modules $\mathfrak{M},\cdots$ are necessary, except that the modules be generated over h by a finite or infinite set of generators M_i such that every $M \in \mathfrak{M}$ is a finite sum

$$M = \sum h_i M_i, \qquad h_i \in h. \tag{1}$$

$\mathfrak{M}$ is called a *free h-module* and M_i a *basis of* $\mathfrak{M}$ if every $M \in \mathfrak{M}$ can be represented uniquely in the way (1).

All considerations can be specialized in the case of graded modules in such a way that all maps which are mentioned in the following are (*homogeneous*) *of degree* 0. It suffices to mention this once and for all.

The superscripts used in the following do not indicate the degree as in §1.

A free resolution of a h-module $\mathfrak{M}$ is an exact sequence of free h-modules $\mathfrak{M}^i$:

$$\cdots \longrightarrow \mathfrak{M}^2 \xrightarrow[\mu_2]{} \mathfrak{M}^1 \xrightarrow[\mu_1]{} \mathfrak{M}^0 \xrightarrow[\mu_0]{} \mathfrak{M} \longrightarrow 0.$$

Such a resolution can be constructed as follows. Let M_i be a system of generators of $\mathfrak{M}$. Between the M_i exist relations $\sum h_i M_i = 0$. These relations form also a h-module, the *1st syzygy module* (Hilbert). Let $\mathfrak{M}^0$ be the free module spanned by as many elements M_i^0 as the M_i. Let $\sum_i h_{ji} M_i = 0$ be a generating system of all relations between the M_i. Then $\mathfrak{M}^1$ is the free module spanned by as many elements M_j^1 as the $\sum_i h_{ji} M_i$. Define the maps μ_1, μ_0 by

$$\mu_1 M_j^1 = \sum_i h_{ji} M_i^0, \qquad \mu_0 M_i^0 = M_i.$$

Then apparently the sequence is exact at $\mathfrak{M}^0$ and $\mathfrak{M}$. This idea can even be generalized as follows:

Proposition 1. *Let an exact sequence*

(2) $$0 \longrightarrow N \xrightarrow[\iota]{} M \xrightarrow[\sigma]{} Q \longrightarrow 0$$

be given. There exist free resolutions of all 3 modules such that rows and columns of the diagram are exact and squares commutative:

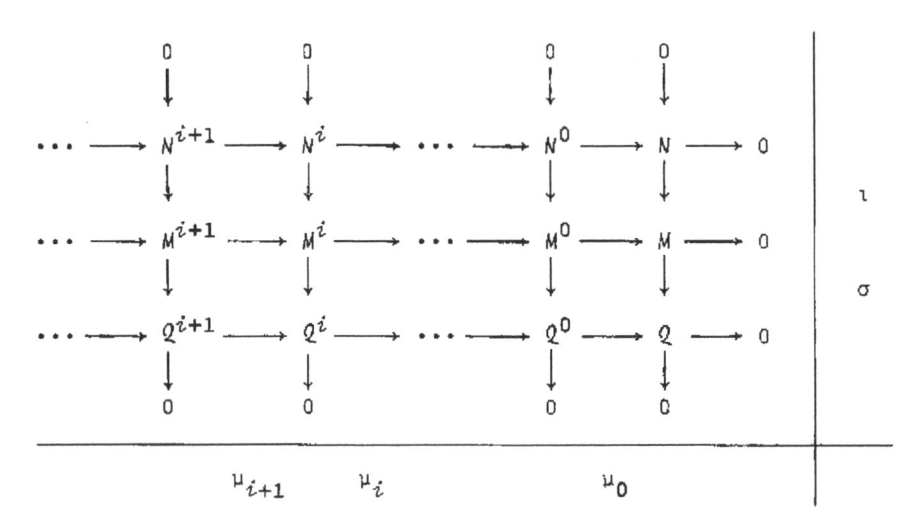

For $i \geq 0$ *we have even*

(3) $$M^i \cong N^i \oplus Q^i .$$

Proof. Let N_ν be a system of generators of N. If the ιN_ν do not yet generate M we add elements $M_\nu \in M$ such that ιN_ν and M_ν together generate M. Then because of the exactness of (2) the σM_ν generate Q. Now we form 3 free modules and 2 maps ι, σ

$$N^0 = \oplus\, hN_\nu^0, \quad M^0 = \oplus\, h\iota N_\nu^0 + hM_\nu^0, \quad Q^0 = \oplus\, h\sigma M_\nu^0$$

and another map μ_0 by

$$\mu_0 N_\nu^0 = N_\nu, \quad \mu_0 \iota N_\nu^0 = \iota\mu_0 N_\nu^0 = \iota N_\nu, \quad \mu_0 M_\nu^0 = M_\nu, \quad \mu_0 \sigma M_\nu^0 = \sigma\mu_0 M_\nu^0 = \sigma M_\nu .$$

This establishes the first 2 columns on the right.

Assume now that N^i, M^i, Q^i are already constructed. The map μ_i leads to some relations

$$(4)\qquad \begin{cases} \sum_{\nu} a_{\mu\nu}\mu_i N_\nu^i = 0, \\ \sum_{\nu} (b_{\mu\nu}\mu_i \iota N_\nu^i + c_{\mu\nu}\mu_i M_\nu^i) = 0, \\ \sum_{\nu} c_{\mu\nu}\mu_i \sigma M_\nu^i = 0 \end{cases}$$

with $a_{\mu\nu},\dots \in h$. Assume that these relations generate the h-modules of all relations between the images of the N_ν^i etc. in N^i etc. Applying ι to the first relations we obtain $\sum a_{\mu\nu}\mu_i \iota N_\nu^i = 0$. We do not include these relations among those of the 2nd kind. Applying σ to the second relations we obtain those of the 3rd kind. We claim that all relations of the 3rd kind are obtained in this way. Indeed consider the element $M^i = \sum_{\nu} d_{\mu\nu}M_\nu^i$ attached to such a relation. Because of $\mu_i\sigma = \sigma\mu_i$, the element $M^{i-1} = \sum_{\nu} d_{\mu\nu}\mu_i M_\nu^i$ of M^{i-1} is mapped on 0 by σ. Because of the exactness of the $(i-1)$th column, $M^{i-1} = \sum_{\nu} b_{\mu\nu}\mu_i \iota N_\nu^i$ or

$$-\sum_{\nu} b_{\mu\nu}\mu_i \iota N_\nu^i + \sum_{\nu} d_{\mu\nu}\mu_i M_\nu^i = 0.$$

We now define 3 new free modules

$$N^{i+1} = \bigoplus hN_\mu^{i+1}, \qquad M^{i+1} = \bigoplus hM_{0,\mu}^{i+1} \bigoplus hM_\mu^{i+1}, \qquad Q^{i+1} = \bigoplus hQ_\mu^{i+1}$$

and a map μ_{i+1} satisfying

$$(5)\qquad \mu_{i+1}N_\mu^{i+1} = \sum_{\nu} a_{\mu\nu}N_\nu^i, \qquad \mu_{i+1}M_{0,\mu}^{i+1} = \sum_{\nu} a_{\mu\nu}\iota N_\nu^i,$$

$$\mu_{i+1}M_\mu^{i+1} = \sum_{\nu} (b_{\mu\nu}\iota N_\nu^i + c_{\mu\nu}M_\nu^i), \qquad \mu_{i+1}Q_\mu^{i+1} = \sum_{\nu} c_{\mu\nu}\sigma M_\nu^i$$

with the same coefficients as in (4). Thus the rows become exact at i. Furthermore we define maps ι, σ by

$$\iota N_\mu^{i+1} = M_{0,\mu}^{i+1}, \qquad \sigma M_{0,\mu}^{i+1} = 0, \qquad \sigma M_\mu^{i+1} = Q_\mu^{i+1}.$$

With this (3) and the exactness of the $(i+1)$th column become evident. Finally $\iota\mu_{i+1} = \mu_{i+1}\iota$ and $\sigma\mu_{i+1} = \mu_{i+1}\sigma$ can immediately be verified.

Proposition 2. Let $\cdots \longrightarrow M_1^0 \longrightarrow M \longrightarrow 0$ *and* $\cdots \longrightarrow M_2^0 \longrightarrow M \longrightarrow 0$ *be two free resolutions of* M. *There exists a third free resolution of* M *with*

$$M^i = M_1^i \oplus M_2^i \oplus N^i \tag{6}$$

where the N^i *are also free modules.*

Proof. Let $M_{1\nu}^0$, $M_{2\nu}^0$ be the bases of M_1^0, M_2^0. We put

$$M^0 = M_1^0 \oplus M_2^0 = \oplus\, hM_{1\nu}^0 \oplus hM_{2\nu}^0$$

and define $\mu_0 : M^0 \longrightarrow M$ in the natural way. In M we have three sets of relations:

$$\sum_\nu a_{\mu\nu}\mu_0 M_{1\nu}^0 = 0, \quad \sum_\nu b_{\mu\nu}\mu_0 M_{2\nu}^0 = 0, \quad \sum_\nu c_{\mu\nu}\mu_0 M_{1\nu}^0 + \sum_\nu d_{\mu\nu}\mu_0 M_{2\nu}^0 = 0 \tag{7}$$

with $a_{\mu\nu}, \cdots \in h$. The first two follow from the two given resolutions; the third set is additional, none of it is free from the $M_{1\nu}^0$ or $M_{2\nu}^0$. We assume that all relations between the $M_{1\nu}^0$, $M_{2\nu}^0$ are consequences of (7). We now introduce elements $M_{1\mu}^1$, $M_{2\mu}^1$, N_μ^1 and form

$$M^1 = \oplus\, hM_{1\mu}^1 \oplus hM_{2\mu}^1 \oplus hN_\mu^1,$$

and we define a map μ_1 by

$$\left\{\begin{aligned} \mu_1 M_{1\mu}^1 &= \sum_\nu a_{\mu\nu} M_{1\nu}^0, \quad \mu_1 M_{2\mu}^1 = \sum_\nu b_{\mu\nu} M_{2\nu}^0, \\ \mu_1 M_\mu^1 &= \sum_\nu c_{\mu\nu} M_{1\nu}^0 + \sum_\nu d_{\mu\nu} M_{2\nu}^0 . \end{aligned}\right. \tag{8}$$

Comparing (7) and (8) we see that

$$M^1 \xrightarrow[\mu_1]{} M^0 \xrightarrow[\mu_0]{} M \longrightarrow 0$$

is exact at M^0,

Now assume that M^i is already constructed, satisfying (6), and such that the sequence is exact in $M, M^0, \cdots, M^{i-1}$. Let $M_{1\nu}^i$, $M_{2\nu}^i$, N_ν^i

be bases of M_1^i, M_2^i, N^i. Then there are relations

$$\text{(9)} \quad \begin{cases} \sum_\nu a_{\mu\nu}\mu_i M_{1\nu}^i = 0, \quad \sum_\nu b_{\mu\nu}\mu_i M_{2\nu}^i = 0, \\ \sum_\nu c_{\mu\nu}\mu_i M_{1\nu}^i + \sum_\nu d_{\mu\nu}\mu_i M_{2\nu}^i + \sum_\nu e_{\mu\nu}\mu_i N_\nu^i = 0. \end{cases}$$

The first 2 sets follow from the given resolutions of M, the third is additional. Accordingly we introduce 3 sets of elements $M_{1\mu}^{i+1}$, $M_{2\mu}^{i+1}$, N_μ^{i+1} which are bases of modules M_1^{i+1}, M_2^{i+1}, N^{i+1}, and map them in the way

$$\text{(10)} \quad \begin{cases} \mu_{i+1}M_{1\mu}^{i+1} = \sum_\nu a_{\mu\nu}M_{1\nu}^i, \quad \mu_{i+1}M_{2\mu}^{i+1} = \sum_\nu b_{\mu\nu}M_{2\nu}^i, \\ \mu_{i+1}N_\mu^i = \sum_\nu c_{\mu\nu}M_{1\nu}^i + \sum_\nu d_{\mu\nu}M_{2\nu}^i + \sum_\nu e_{\mu\nu}N_\nu^i \end{cases}$$

on M^i. Comparison between (9) and (10) yields the exactness at M^i.

§3. *The functors* $\mathrm{Ext}_h^i(M,H)$

Two h-modules M, H and a free resolution

$$\text{(1)} \quad \cdots \longrightarrow M^2 \xrightarrow[\mu_2]{} M^1 \xrightarrow[\mu_1]{} M^0 \xrightarrow[\mu_0]{} M \longrightarrow 0$$

of M are given. (1) entails another sequence

$$\text{(2)} \quad \cdots \xleftarrow[\mu_2^*]{} \mathrm{Hom}_h(M^1,H) \xleftarrow[\mu_1^*]{} \mathrm{Hom}_h(M^0,H) \xleftarrow[\mu_0^*]{} 0$$

($\mathrm{Hom}_h(M,H)$ is left out). The map μ_0^* is trivial. For the definition of μ_i^* we denote elements of $\mathrm{Hom}_h(M^i,H)$ by M^{i*}; in this way a product $M^{i*}M^i$ with a $M^i \in M^i$ is an element of H. Now μ_i^* is defined as the right operator on M^{i*} defined by

$$\text{(3)} \quad (M^{i-1,*}\mu_i^*)M^i = M^{i-1,*}(\mu_i M^i).$$

By (3) $M^{i-1,*}\mu_i^*$ becomes an element of $\mathrm{Hom}_h(M^i,H)$. Because of $\mu_{i-1}\mu_i = 0$ we have

$$(M^{i-2,*}\mu_{i-1}^*\mu_i^*)M^i = (M^{i-2,*}\mu_{i-1}^*)(\mu_i M^i)$$
$$= M^{i-2,*}(\mu_{i-1}\mu_i M^i) = 0,$$

therefore $\mu_{i-1}^*\mu_i^* = 0$, and (2) is a complex. But (2) need not be exact.

We introduce the *homology groups* of (2):

$$\mathrm{Ext}_h^i(M,N) = \ker(\mu_{i+1}^*)/\mathrm{im}(\mu_i^*). \tag{4}$$

They are h-modules (right modules if h is not commutative) and even graded modules if h, M, N are graded.

Proposition 1. $\mathrm{Ext}_h^0(M,H) = \mathrm{Hom}_h(M,H)$.

Proof. Define a map $\mu^*:\mathrm{Hom}_h(M,H) \longrightarrow \mathrm{Hom}_h(M^0,H)$ by

$$(M^*\mu^*)M^0 = M^*(\mu_0 M^0).$$

Because μ_0 in (1) is surjective, $M^*\mu^* = 0$ if and only if $M^* = 0$. Therefore μ^* is injective. What we have to prove is the following: if for a $M^{0*} \in \mathrm{Hom}_h(M^0,H)$ $M^{0*}\mu_1^* = 0$, then there exists a $M^* \in \mathrm{Hom}_h(M,H)$ such that $M^{0*} = M^*\mu^*$.

The assumption entails $M^{0*}(\mu_1 M^1) = 0$ for all $M^1 \in M^1$. Therefore M^{0*} acts on $M^0/\mathrm{im}\ \mu^1$ which is isomorphic with M by (1). So there exists an M^* such that $M^{0*} - M^*\mu^*$ is the zero operator on $M^0/\mathrm{im}\ \mu^1$. It is then also the zero operator on M^0. QED.

Theorem 1. *To a "short" exact sequence*

$$0 \longrightarrow N \xrightarrow[\iota]{} M \xrightarrow[\sigma]{} Q \longrightarrow 0 \tag{5}$$

there exist 2 "long" exact sequences ($\mathrm{Ext}^0 = \mathrm{Hom}$)

$$0 \longrightarrow \mathrm{Ext}_h^0(Q,H) \longrightarrow \cdots \longrightarrow \mathrm{Ext}_h^i(Q,H) \xrightarrow[\sigma]{} \mathrm{Ext}_h^i(M,H) \tag{6}$$
$$\xrightarrow[\iota]{} \mathrm{Ext}_h^i(N,H) \xrightarrow[\tau]{} \mathrm{Ext}_h^{i+1}(Q,H) \longrightarrow \cdots$$

(7)
$$0 \longrightarrow \mathrm{Ext}_h^0(H,N) \longrightarrow \cdots \longrightarrow \mathrm{Ext}_h^i(H,N) \xrightarrow[\iota]{} \mathrm{Ext}_h^i(H,M)$$
$$\xrightarrow[\sigma]{} \mathrm{Ext}_h^i(H,Q) \xrightarrow[\tau]{} \mathrm{Ext}_h^{i+1}(H,N) \longrightarrow \cdots .$$

The maps ι, σ are easy consequences of those in (5), τ is to be defined.

If h, M, N are graded and ι, σ of degree 0, then also τ is of degree 0.

The proof will follow from another theorem which we have to prove first.

Theorem 2. Assume a diagram

$$\begin{array}{ccccccccc|c}
 & & 0 & & 0 & & 0 & & & \\
 & & \uparrow & & \uparrow & & \uparrow & & & \\
\cdots & \longleftarrow & N^{i+1} & \longleftarrow & N^i & \longleftarrow \cdots \longleftarrow & N^0 & \longleftarrow & 0 & \\
 & & \uparrow & & \uparrow & & \uparrow & & & \iota \\
\cdots & \longleftarrow & M^{i+1} & \longleftarrow & M^i & \longleftarrow \cdots \longleftarrow & M^0 & \longleftarrow & 0 & \\
 & & \uparrow & & \uparrow & & \uparrow & & & \sigma \\
\cdots & \longleftarrow & Q^{i+1} & \longleftarrow & Q^i & \longleftarrow \cdots \longleftarrow & Q^0 & \longleftarrow & 0 & \\
 & & \uparrow & & \uparrow & & \uparrow & & & \\
 & & 0 & & 0 & & 0 & & & \\
\hline
 & & \mu_{i+1} & & \mu_i & & \mu_0 & & &
\end{array}$$

in which the rows are complexes (i.e., $\mu_{i+1}\mu_i = 0$), the columns are exact sequences, and the squares are commutative. Then the "homology groups"

(8)
$$H^i(N) \quad = \quad \ker \mu_{i+1} / \operatorname{im} \mu_i \qquad \textit{in } N^i, \textit{ etc.}$$

form an exact sequence

(9)
$$0 \longrightarrow H^0(Q) \longrightarrow \cdots \longrightarrow H^i(Q) \xrightarrow[\sigma]{} H^i(M)$$
$$\xrightarrow[\iota]{} H^i(N) \xrightarrow[\tau]{} H^{i+1}(Q) \longrightarrow \cdots ,$$

where the maps σ, ι are defined in a natural way by the given diagram and τ has yet to be defined.

If h, M, N, Q are graded and ι, σ of degree 0, τ has also degree 0.

Proof. The maps ι, σ act as well on the $\operatorname{im}\mu_i$ and $\ker\mu_{i+1}$ because the squares are commutative. So they also act on the homology groups, and $\sigma\iota$ remains 0. In the following we denote elements of M^i etc. by the corresponding Latin letter. The maps will be written as right operators.

1) For a $M^i \in \ker\mu_{i+1}$ assume $M^i\iota = N^{i-1}\mu_i$. Because of the exactness of the columns $N^{i-1} = M^{i-1}\iota$ with some M^{i-1}, and now $(M^i - M^{i-1}\mu_i)\iota = 0$. Therefore $M^i - M^{i-1}\mu_i = Q^i\sigma$. This entails that, as an element of $H^i(M)$, M^i is an image of an element of $H^i(\mathcal{Q})$. In other words, (9) is exact at $H^i(M)$.

2) Construction of τ: We start from a N^i with $N^i\mu_{i+1} = 0$. Let $N^i = M^i\iota$ and therefore $M^i\iota\mu_{i+1} = M^i\mu_{i+1}\iota = 0$. Because of the exactness of the columns $M^i\mu_{i+1} = Q^{i+1}\sigma$. Define

$$N^i\tau = Q^{i+1}. \tag{10}$$

In this construction M^i is not uniquely determined by N^i, but only up to an M_0^i with $M_0^i\iota = 0$. This means $M_0^i = Q^i\sigma$ and $M_0^i\mu_{i+1} = Q^i\sigma\mu_{i+1} = Q^i\mu_{i+1}\sigma = Q_0^{i+1}\sigma$. Therefore τ maps N^i on an element of $H^{i+1}(\mathcal{Q})$.

If $N^i = N^{i-1}\mu_i$ and $N^{i-1} = M^{i-1}\iota$, then, in the above construction, $Q^{i+1} = 0$. Therefore τ is a homomorphism

$$\tau : H^i(N) \longrightarrow H^{i+1}(\mathcal{Q}).$$

3) The product map yields $N^i\tau\sigma = Q^{i+1}\sigma = M^i\mu_{i+1}$ which is zero in $H^{i+1}(M)$.

Conversely, consider Q^{i+1} with $Q^{i+1}\sigma = M^i\mu_{i+1}$. Then $Q^{i+1}\sigma\iota = M^i\mu_{i+1}\iota = M^i\iota\mu_{i+1} = N^i\mu_{i+1} = 0$, and $Q^{i+1} = N^i\tau$. Thus (9) is exact at $H^{i+1}(\mathcal{Q})$.

4) If, in the construction of τ, $N^i = M^i\iota$ with an $M^i \in H^i(M)$ or, in other words, with $M^i\mu_{i+1} = 0$, we see $Q^{i+1} = 0$. This shows

$\iota\tau = 0$.

Conversely, if $N^i\tau = 0$, we have $N^i = M^i\iota$ with $M^i\mu_{i+1} = 0$. So (9) is exact at $H^i(N)$.

5) The exactness at $H^0(Q)$ remains to be proved. This means that $Q^0\sigma = 0$ if and only if $Q^0 = 0$ which follows from the exactness of the last column in the given diagram.

Proof of Theorem 1. We construct a family of free resolutions N^i, M^i, Q^i of N, M, Q according to Proposition 1 in §2. Then the $\mathrm{Hom}_h(N^i,H)$, etc. form a diagram of the sort considered in Theorem 2. (The exactness of the columns follows from §2, (3).) So we replace N^i etc. in Theorem 2 by $\mathrm{Hom}_h(N^i,H)$ etc. and obtain by comparison of (4) and (8): $\mathrm{Ext}_h^i(N,H)$ etc. instead of $H^i(N)$ etc. Then (9) becomes (6).

For the proof of (7) we take a free resolution H^i of H and form the diagram

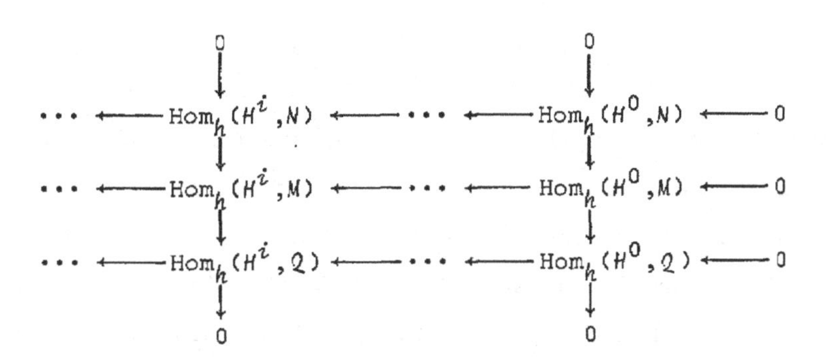

First we remark that the columns are exact. For this we use that the H^i are free, i.e., direct sums of modules hH. Because of §1, Theorem 1, the $\mathrm{Hom}_h(H^i,N)$ etc. become direct sums of $\mathrm{Hom}_h(hH,N)$ etc., and these modules are isomorphic with N etc.

Secondly we see that this diagram is commutative. Indeed denote the homomorphism $H^i \longrightarrow H^{i-1}$ by κ_i, and let

$$\phi_i : H^i \longrightarrow N$$

be an element of $\mathrm{Hom}_h(H^i, N)$. Then

$$\iota(\phi_i \kappa_{i+1}) = (\iota\phi_i)\kappa_{i+1} : H^{i+1} \longrightarrow \iota N.$$

The commutativity of our diagram simply follows from the fact that κ_{i+1} and ι act on 2 different modules. The same argument applies for the other squares.

Lastly $\kappa_i \kappa_{i+1} = 0$, which is evident.

Now we obtain (7) from (9).

§4. *The functors* $\mathrm{Ext}_h^i(M,N)$ *continued*

Theorem 1. *The h-modules* $\mathrm{Ext}_h^i(M,N)$ *are independent of the free resolution* M^i *of* M *used in their definition.*

Proof. We begin with discussing a free resolution of the zero module, that is an exact sequence of free modules

$$\cdots \longrightarrow M^2 \xrightarrow[\mu_2]{} M^1 \xrightarrow[\mu_1]{} M^0 \xrightarrow[\mu_0]{} 0. \tag{1}$$

We contend that all $\mathrm{Ext}_h^i(M,N) = 0$ for an arbitrary module N. We show this by suitably altering this resolution without changing the Ext^i. Let M_ν^0 be a basis of M^0 and M_ν^1 elements of M^1 which are mapped on the M_ν^0 by μ_1. They form a free submodule $M_1^1 \subseteq M^1$, isomorphic with M^0.

Denote the kernel of μ_1 in M^1 by M_0^1. Then $M_0^1 \cap M_1^1 = 0$ and thus

$$M^1 = M_1^1 \oplus M_0^1. \tag{2}$$

Now we take 2 other modules $\tilde{M}_1^1$ and $\tilde{M}^2$, both isomorphic with M_1^1 and form a new exact sequence

$$\cdots \xrightarrow[\tilde{\mu}_3]{} M^2 \oplus \tilde{M}^2 \xrightarrow[\tilde{\mu}_2]{} M^1 \oplus \tilde{M}_1^1 \xrightarrow[\tilde{\mu}_1]{} M^0 \longrightarrow 0, \tag{3}$$

the maps $\tilde{\mu}_1, \cdots$ defined as follows:

$$\tilde{\mu}_1 M^1 = \mu_1 M^1 , \qquad \tilde{\mu}_1 \tilde{M}^1_1 = 0 ,$$

$$\tilde{\mu}_2 M^2 = \mu_2 M^2 , \qquad \tilde{\mu}_2 \tilde{M}^2 = \tilde{M}^1_1 ,$$

$$\tilde{\mu}_3 M^3 = \mu_3 M^3 .$$

urthermore

$$M^1 \oplus \tilde{M}^1_1 = M^1_1 \oplus (M^1_0 \oplus \tilde{M}^1_1) = M^1_1 \oplus M^1_2 ,$$

nd, because of (2), M^1_2 is free. Also $M^2 \oplus \tilde{M}^2$ is the direct sum of 2 ree modules and therefore free. This (3) is an exact sequence of free odules in which the second module is the direct sum of two free modules ne of which is mapped isomorphically on the last, and the other on zero.

As is easily checked, the homology groups of the Hom's of (1) and 3) are the same.

We write again M^i for the members of (3). In a second step we lter M^2, M^3 similarly by adding modules $\tilde{M}^2_1 \cong \tilde{M}^3 \cong M^1_2$ which are mapped n 0 and $\tilde{M}^2_1$ respectively. The homology groups remain again unchanged.

At length we arrive at an exact sequence (1) in which

$$M^i = M^i_1 \oplus M^i_2 , \qquad \mu_i M^i_1 = M^{i-1}_2 , \qquad \mu_i M^i_2 = 0 .$$

or this all homology groups of the Hom's vanish.

After this preparation we can proceed with the proof of Theorem 1. ake 2 free resolutions M^i_1, M^i_2 of M and form a third free resolution M^i_3 ccording to Proposition 2 of §2. We shall show the Ext for M^i_1 and $^i_3$ are equal. A symmetry argument then yields the equality of the Ext or M^i_1 and M^i_2. But we change the notation, writing M^i_2 instead of M^i_3. hen we assume that

$$M^i_2 = M^i_1 \oplus N^i \tag{4}$$

ith free modules N^i.

Now we form the diagram

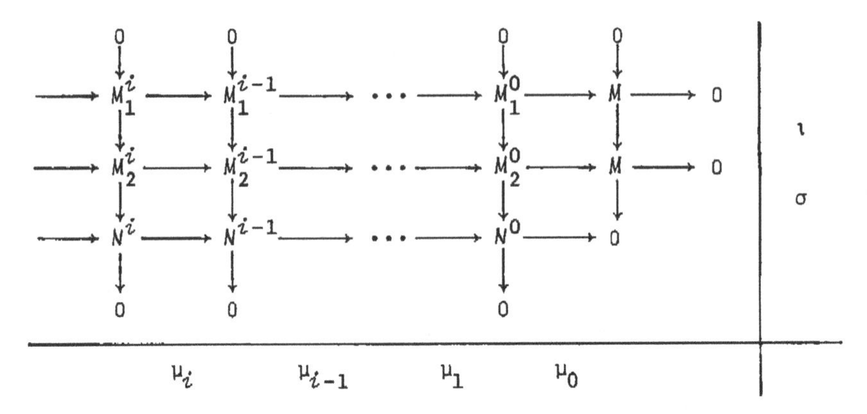

The columns are exact. The first two rows are given by the 2 free resolutions of M; they are exact sequences. The maps in the last row have yet to be defined. The upper squares are commutative.

Because of (4), the $N^i \subseteq M_2^i$. Therefore $\mu_i N^i$ is defined, and again due to (4):

$$\mu_i N^i = M_1^{i-1} + N^{i-1} \qquad (M_1^{i-1} \in M_1^{i-1})$$

where both terms are uniquely determined. We define

$$\mu_i^! N^i = N^{i-1}$$

and use these $\mu_i^!$ in the lower row of our diagram. With this definition also the lower squares become commutative.

Now $\mu_i \mu_{i+1} N^{i+1} = \mu_i M_1^i + \mu_i N^i = \mu_i M_1^i + M_1^{i-1} + N^{i-1} = 0$. Because of (4) therefore $N^{i-1} = 0$. Also

$$\mu_i^! \mu_{i+1}^! N^{i+1} = \mu_i^! N^i = N^{i-1} = 0.$$

Thus the lower row is a complex.

We want to show that the lower row is even exact. Assume $\mu_i^! N^i = 0$. N^i can be considered as a σM_2^i and then $\mu_i M_2^i = M_1^{i-1} \in M_1^{i-1}$ with $\mu_{i-1} M^{i-1} = 0$. Because of the exactness of the upper row $M_1^{i-1} = \mu_i M_1^i$, and then $\mu_i (M_2^i - \iota M_1^i) = 0$, $M_2^i - \iota M_1^i = \mu_{i+1} M_2^{i+1}$ because of the exactness of the middle row. Now

$$N^i = \sigma(M_2^i - \iota M_1^i) = \sigma\mu_{i+1}M_2^{i+1} = \mu_{i+1}\sigma M_2^{i+1} = \mu_{i+1}N^{i+1}.$$

Since $N^i \in N^i$, $\mu_{i+1}N^{i+1} = \mu'_{i+1}N^{i+1}$. This shows the exactness.

At last we apply Theorem 2 of the preceding section to our diagram and use the fact that the homology groups of the last row are 0. Distinguishing the Ext formed with the first 2 rows by a subscript we thus obtain the exact sequence

$$\cdots \longrightarrow 0 \longrightarrow \mathrm{Ext}_h^i(M,N)_2 \longrightarrow \mathrm{Ext}_h^i(M,N)_1 \longrightarrow 0 \cdots,$$

whence

$$\mathrm{Ext}_h^i(M,N)_1 \cong \mathrm{Ext}_h^i(M,N)_2.$$

An immediate consequence of §1, Theorem 1 and of the definition of the Ext is

Theorem 2. $\mathrm{Ext}_h^i(M_1 \oplus M_2, N) = \mathrm{Ext}_h^i(M_1,N) + \mathrm{Ext}_h^i(M_2,N)$.

From Theorem 1 and Theorem 2 we derive

Theorem 3. *If* M *is a direct summand of a free module, all* $\mathrm{Ext}_h^i(M,N)$ *vanish.*

Indeed for a free M the $\mathrm{Ext}_h^i(M,N)$ are the same as the $\mathrm{Ext}_h^{i+1}(0,N)$ for the zero-module which are 0. If $M_1 = M \oplus M_2$ with M_1 free, $\mathrm{Ext}_h^i(M,N) = 0$ because of Theorem 2.

Remark. Modules which are direct summands of free modules are called *projective modules*. One can replace free resolutions by "projective resolutions," and the appropriate homology groups are also equal to the Ext. But in the applications which we have in mind projective modules are always free (§8), and the generality does not pay.

§5. *Modules over polynomial rings*

From now on

$$h = h_n = k_0[y_0, \cdots, y_n]$$

is the graded polynomial ring in $n+1$ indeterminates y_ν over a field k_0. The subscript n will in general be omitted. We repeat what we said in §1 that we only consider homogeneous elements, but we do not mention this again. In other words, all elements of h are homogeneous. The same applies to h-modules.

We put

$$x_i = \frac{y_i}{y_0}, \quad k = k_n = k_0(x_1, \cdots, x_n),$$

the field of rational functions over k_0. All quotients of elements of h form the graded division ring

$$\overline{k} = \{\tfrac{a}{b}:\ a, b \in h,\ b \neq 0\} = \bigoplus_{i=-\infty}^{+\infty} k y_0^i .$$

Let R/k be the N-dimensional vector space over k. We also intro duce

$$\overline{R} = \overline{k}R = \{\tfrac{r}{b}:\ r \in R,\ b \neq 0 \in h\}.$$

A torsion free h-module M can always be embedded in the vector space $\overline{R} = \overline{k}M$ over $\overline{k}$. Evidently $\overline{R}$ is generated by a space R over k whose vectors have degrees 0.

To each prime polynomial $p \in h$ there corresponds the p-adic valuation of h and $\overline{k}$. The valuation ring is

$$h_p = \{\tfrac{a}{b}:\ a, b \in h;\ b \not\equiv 0 \bmod p\}.$$

To each torsion free module M we introduce its p-adic extension

$$M_p = h_p M.$$

Proposition 1. *A finite torsion-free h_p-module M_p is a free h_p-module and possesses a basis of elements of degree* 0.

Proof. Form the subring h_p^0 and submodule M_p^0 of elements of degree

. h_p^0 is a principal ideal ring and M_p^0 a finite h_p^0-module. It has a asis M_ν^0. These M_ν^0 evidently also form a basis of M_p.

Let $\bar{R}^*$ be the dual space to $\bar{R} = \bar{k}M$. We introduce the dual module or complement of M:

$$M^* = \mathrm{Hom}_h(M,h) = \{M^* \in \bar{R}^* : M^*M \subseteq h\}.$$

vidently

$$\text{1)} \qquad M^{**} \supseteq M \quad \text{and} \quad M^{***} = M^*.$$

f $M^{**} = M$, M is called *reflexive*. Reflexive modules will play an mportant rôle throughout.

Theorem 1. *The following* 3 *conditions on a finite torsion-free odule* M *are equivalent.*

a) $M^{**} = M$,

b) if $M \in \bar{R}$ *and with relatively prime* $c_i \in h$: $c_iM \in M$, *then* $M \in M$.

c) $M = \bigcap_p M_p$.

Corollary. A free module is reflexive, and the complement of a odule is reflexive (because of the second equation (1)).

Proof. $a \Rightarrow b$) Assume M satisfies the conditions under b). Then $_iM^* \subseteq h$, and because the c_i are relatively prime, $M^*M \subseteq h$. Therefore $\in M^{**}$ and because of the assumption $M \in M$.

$b \Rightarrow c$) Assume $M \in \bigcap_p M_p$. Let M_ν be a system of generators of M. s $M \in M_p$ for each p, $M = \sum h_\nu M_\nu$ with p-adic integers h_ν. Thereore there exists a $c_p \in h$, prime to p, such that $c_p h_\nu \in h$. Then $_pM \in M$. Since all c_p have no common divisor, the condition under b) s satisfied, and so $M \in M$.

Thus we have proved $M \supseteq \bigcap_p M_p$. The opposite inclusion is trivial.

$c \Rightarrow a$) We introduce the complements of the p-adic extensions

$$M_p{}^* = \{M^* \in \bar{R}^*: M^*M_p \subseteq h_p\}.$$

Evidently

$$M_p{}^* = M^*{}_p,$$

in other words the formations of the complement and of the p-adic extension commute. For $M_p = \bigoplus h_p M_\nu$ (Proposition 1) the complement is $M_p{}^* = \bigoplus h_p M_\nu^*$ with $M_\mu^* M_\nu = \begin{cases} 1 \text{ for } \mu = \nu, \\ 0 \text{ for } \mu \neq \nu. \end{cases}$ This entails $M_p{}^{**} = M_p$.

Now from $M = \bigcap_p M_p$ follows:

$$M^* = \{M^* \in \bar{R}^*: M^*M_p \subseteq h_p \text{ for all } p\}.$$

This means

$$M^* = \bigcap_p M_p^*.$$

The reasoning can be repeated:

$$M^{**} = \bigcap_p M_p{}^{**} = \bigcap_p M_p = M, \qquad \text{QED.}$$

A graded free h-module M has always a basis of homogeneous elements. This can easily be proved. However we need not show this if we make the convention that a free module is always

$$M = \bigoplus hM_\nu$$

with homogeneous $M_\nu \in M$. But here we can even omit "homogeneous" referring to the convention made at the beginning of this section that non-homogeneous elements do not exist for us.

In what follows a polynomial $f \in h$ will be called *normed* if it has the form

$$f = y_n^m + a_1(y)y_n^{m-1} + \dots + a_m(y), \tag{2}$$

with $c_i(y)$ polynomials of degrees i in $y_0,\cdots,y_{n-1}$. If k_0 has ∞ many elements one can transform the variables y_ν in such a way that a given polynomial is normed (even that finitely many given polynomials are normed).

Theorem 2. *Let M, N be h-modules in $\overline{R}$, $N \subseteq M$ and with a normed polynomial $f: fM \subseteq N$. If N has property b) of Theorem* 1, *the quotient module M/N is a torsion-free module with respect to*

$$h_{n-1} = k_0[y_0,\cdots,y_{n-1}]. \tag{3}$$

Conversely, if M/N is a torsion free h_{n-1}-module and if M has property b) of Theorem 1, *then N has also this property.*

Proof. That M/N is a h_{n-1}-module is evident. Assume $h\overline{M} = 0$ for $h \in h_{n-1}$ and $\overline{M} \in M/N$.

Take an $M \in M$ whose residue in M/N is $\overline{M}$. Then $hM \in N$. On the other hand $fM \in N$ with f of the form (2). h and f are relatively prime and because of the assumption and Theorem 1: $M \in N$. Therefore $\overline{M} = 0$.

Conversely let $M \in \overline{R}$ and $c_i \in h$ and relatively prime such that $c_iM \in N$. Then also $c_iM \in M$ and, because M is supposed reflexive, $M \in M$. There exist polynomials $u_i \in h$ such that $h = \sum u_ic_i$ does not contain y_n. Now $hM \in N$, and for the residue mod N we have $h\overline{M} = 0$. Because M/N is torsion free, $\overline{M} = 0$ or $M \in N$. Thus, by Theorem 1, N is reflexive.

Corollary. If M and N are finite modules of equal rank $N \subseteq M$, and if k_0 has ∞ many elements, then there is, after a suitable linear transformation of the variables y_ν, a normed polynomial f such that $fM \subseteq N$, and Theorem 2 *can be applied.*

Namely let M_ν be a system of generators of M. For each ν there is an $f_\nu \in h$ such that $f_\nu M_\nu \in N$. Then $fM \subseteq N$ with $f = \prod f_\nu$. Now one can proceed as said above.

§6. *The rank polynomial*

We denote the number of (with respect to k_0) linearly independent elements of degree λ of a h-module M by $L(\lambda,M)$ and show

Proposition 1. *If M is a finite h-module $L(\lambda,M)$ coincides for sufficiently large λ with a polynomial*

$$H(\lambda,M) = \gamma_n(M)\binom{\lambda}{n} + \gamma_{n-1}(M)\binom{\lambda}{n-1} + \cdots + \gamma_0(M)$$

where the $\gamma_\nu(M)$ are integers.

$L(\lambda,M)$ and $H(\lambda,M)$ are class invariants; that means they are the same for M and mM if $m \in k$.

$H(\lambda,M)$ is called the *rank polynomial* or the *Hilbert polynomial* of M.

Proof by induction on n. For $n = 0$ we proceed by induction on the minimal number of generators of M. An element $M \in$ M of smallest degree is obviously contained in any set of generators. The proposition is true for the submodule hM, $H(\lambda,hM)$ being either 0 or 1 according as M is a torsion element or not. Now by (1) below

$$L(\lambda,M) = L(\lambda,hM) + L(\lambda,M/hM).$$

M/hM has less generators than M. So both terms on the right become constants for sufficiently large λ, and therefore also $L(\lambda,M)$. Denote the ranks of the M_1^i by γ_i. Then for sufficiently large λ: $L(\lambda,M) = \sum \gamma_i = \gamma_0(M)$.

For the induction, and also for later purposes we need

$$(1) \qquad L(\lambda,M/N) = L(\lambda,M) - L(\lambda,N)$$

if $N \subseteq M$. This is evident. As well evident is

2) $$L(\lambda+\mu,M) = L(\lambda,y_n^{-\mu}M).$$

Assume now the proposition proved for n-1 and put

$$M' = y_n^{-1}M/M$$

nstead of M/N in (1). Then (1) and (2) yield

3) $$L(\lambda,M') = L(\lambda+1,M) - L(\lambda,M).$$

' is a finite h_{n-1}-module (h_{n-1} as in §5, Theorem 2). Therefore for arge λ:

$$L(\lambda,M') = \gamma_{n-1}(M')\binom{\lambda}{n-1} + \cdots + \gamma_0(M')$$

ith some integers $\gamma_\nu(M')$. (3) is a difference equation for $L(\lambda,M)$; its olution is uniquely determined as

$$L(\lambda,M) = \gamma_{n-1}(M')\binom{\lambda}{n} + \cdots + \gamma_0(M')\binom{\lambda}{1} + \gamma_0(M)$$

p to a constant term $\gamma_0(M)$. QED.

The constant term of the rank polynomial will be written without he subscript:

4) $$\gamma(M) = \gamma_0(M).$$

t will be called the *genus coefficient*. From (1) and Proposition 1 we erive

5) $$H(\lambda,M/N) = H(\lambda,M) - H(\lambda,N)$$

nd especially

6) $$\gamma(M/N) = \gamma(M) - \gamma(N).$$

Proposition 2. *If M is a finite torsion-free module of rank N (i.e.,* $= \bar{k}M$ *has rank N over* $\bar{k}$*), the first terms of the rank polynomial are*

$$\gamma_n(M) = N, \qquad \gamma_{n-1}(M) = nN - G(M)$$

where $G(M)$ *is the "linear degree" of* M, *defined as follows:*

Let R_ν^0 be a basis of $\bar{R}/\bar{k}$ having degrees 0 and M_ν^0 a basis of M_p/h_p also having degrees 0 (see Proposition 1 in §5). Then

$$M_\mu^0 = \sum_\nu a_{\mu\nu} R_\nu^0, \qquad a_{\mu\nu} \in k.$$

Let $p^{\nu_p(M)}$ be the exact power of p dividing the determinant $|a_{\mu\nu}|$. Then

$$G(M) = \sum_p \deg(p)\, \nu_p(M). \tag{7}$$

$G(M)$ does not depend on the bases R_ν^0, M_ν^0.

Proof. If M is torsion free and of rank N, the module $M' = y_n^{-1}M/M$ used in the proof of Proposition 1 is also torsion free and has the same rank. Therefore the proof of Proposition 1 yields at the same time the first statement.

The second statement is true for a free module

$$N = hM_1 \oplus \cdots \oplus hM_N$$

with elements M_ν of degrees μ_ν. In this case the rank polynomial is

$$H(\lambda,N) = \sum_{\nu=1}^{N} \binom{\lambda-\mu_\nu+n}{n} = \frac{N}{n!}\lambda^n + \sum_{\nu=1}^{N} \frac{1}{(n-1)!}\left(\frac{n+1}{2} - \mu_\nu\right)\lambda^{\nu-1} + \cdots$$

$$= N\binom{\lambda}{n} + (nN - \sum_{\nu=1}^{N} \mu_\nu)\binom{\lambda}{n-1} + \cdots$$

and

$$G(N) = \sum_{\nu=1}^{N} \mu_\nu = \sum_{\nu=1}^{N} G(hM_\nu).$$

In the general case we proceed by induction on n. The beginning of the induction at $n = 1$ and the induction proper are treated by the same arguments. Take N linearly independent $M_\nu \in M$ and form the free submodule $N = \oplus hM_\nu$. The factor module $M' = M/N$ is torsion free by

5, Theorem 2 and Corollary Theorem 1. The first of our statements for -1 is: $\gamma_{n-1}(M') = N'$ is equal to the rank of M'. Because of (5)

$$\gamma_{n-1}(M) = nN - G(N) + N',$$

nd we have to prove

$$G(N) - G(M) = N'. \tag{8}$$

Because an arbitrary extension of the field k_0 does not change the tatements we may assume k_0 as infinite. Let $fM \subseteq N$ with a polynomial $\in h$, and transform the variables y_ν in such a way that f is normed in he sense of §5, (2). Now

$$M_p = N_p \quad \text{for all } p \nmid f, \tag{9}$$

specially for such p which are not normed in the same sense as f. herefore we obtain the space in which $M' = M/N$ is embedded thus:

$$\bar{R}' = \bar{k}_{n-1}M'$$

ith

$$\bar{k}_{n-1} = \{\tfrac{a}{b} : a,b \in k_0[y_0,\cdots,y_{n-1}]\}.$$

he dimension of $\bar{R}'$ is computed as follows: introduce the principal deal domain

$$k' = k_{n-1}[x_n] = k_0(x_0,\cdots,x_{n-1})[x_n]$$

nd the submodules M^0, N^0 of $k'M$, $k'N$ of elements of degrees 0. They re finite k'-modules and have special bases (elementary divisor theoem)

$$M^0 = \bigoplus k'N_\nu^0, \qquad N^0 = \bigoplus k'f_\nu N_\nu^0 \tag{10}$$

ith polynomials in x_n:

$$f_\nu = x_n^{m_\nu} + c_1(x_1,\dots,x_{n-1})x_n^{m_\nu - 1} + \dots .$$

This shows

$$\bar{R}' = \bigoplus k'/k'f_\nu ,$$

and therefore

(11) $$\dim \bar{R}' = N' = \sum m_\nu .$$

Because of $fM \subseteq N$ and (10) the f_ν divide $y_0^{-\deg f}f$, which is a polynomial in all x_i. Because of the lemma of Gauss the f_ν are also polynomials in all x_i, and the $y_0^{m_\nu} f_\nu \in h$. On the other hand,

$$G(M) - G(N) = \sum_{p/f} \deg(p)\,(\nu_p(M) - \nu_p(N)),$$

and due to (10) this is $= \deg(\prod f_\nu) = \sum m_\nu$. Comparison with (11) yields the wanted result (8).

Proposition 3. *Dual modules have opposite linear degrees.*

Proof: Obvious.

§7. *Reduction of the number of variables*

In this section we take up the situation given in §5, Theorem 2: M and N are finite torsion free modules of equal rank, $M \supseteq N$, and $fM \subseteq N$ with a normed polynomial f. Thus, as a h-module

$$Q = M/N, \qquad fQ = 0$$

is a finite torsion module, but as a h_{n-1}-module, Q may again be torsion free. Our task is to connect properties of Q as a h_{n-1}-module with properties of M and N as h-modules.

For some applications a greater generality is useful. Consider the h-ideal

(1) $$p^r = hy_{n-r+1} + \cdots + hy_n,$$

generated by the last r of the variables y_ν, and form

(2) $$h^p = h^p_r = \{ \tfrac{a}{b} : \ a, b \text{ homogeneous}, \ b \not\equiv 0 \bmod p^r \}.$$

This is a special *local ring*. Evidently h^{r-1}_{n-1} is a similar ring, and *its elements represent the residues of* $h \bmod h^r y_n$ *in a unique way*. Because of this fact h can be replaced throughout § 7 by h^r. But for the sake of simplicity we shall only write h.

Reduction Lemma. Assume $n > 0$. *Let* Q *be a* h*-module with* $fQ = 0$ *where* f *is a normed polynomial:*

$$f = y_n^m + c(y_0, \cdots, y_{n-1}) y_n^{m-1} + \cdots + c_m(y_0, \cdots, y_{n-1}).$$

There exist homogeneous h_{n-1}*-isomorphisms* ρ_i *of degree* 0 *which map (onto)*

$$\rho_i : \mathrm{Ext}^i_h(Q, h) \longrightarrow \eta^{-1} \mathrm{Ext}^{i-1}_{h_{n-1}}(Q, h_{n-1}),$$

where η^{-1} *denotes a homogeneous isomorphism of degree* -1.

The reduction lemma is implicitly contained in *M. Eichler*: Eine Theorie der linearen Räume über rationalen Funktionenkörpern und der Riemann-Rochsche Satz in algebraischen Funktionenkörpern, Math. Annalen 156(1964), 347-377 (Satz 1 and Satz 2). The proof, not using homological algebra was tedious. The first formulation in the present form, together with a complete proof, was given by *P. Roquette* (not published). Another proof by *K. Kiyek* will appear elsewhere.

Proof. We consider the ring q_n of all formal power series in $\frac{1}{y_n}$:

$$c_m y_n^m + c_{m-1} y_n^{m-1} + \cdots$$

with $c_\mu \in h_{n-1}$ in which all terms $c_\mu y_n^\mu$ are homogeneous of the same degrees. With this graded ring we form the short exact sequence

$$0 \longrightarrow h \longrightarrow q_n \longrightarrow q_n/h \longrightarrow 0$$

of maps of degree 0 and the attached series

$$(3) \qquad 0 \longrightarrow \mathrm{Hom}_h(Q,h) \longrightarrow \mathrm{Hom}_h(Q,q_n) \longrightarrow \mathrm{Hom}_h(Q,q_n/h) \longrightarrow \mathrm{Ext}_h^1(Q,h) \longrightarrow \cdots$$

Because f is normed, f^{-1} is a unit in q_n, and therefore

$$\mathrm{Ext}_h^i(Q,q_n) = \mathrm{Ext}_{q_n}^i(q_nQ,q_n) = \mathrm{Ext}_{q_n}^i(q_nfQ,q_n) = 0.$$

Together with (3) this leads to the isomorphism of degree 0

$$(4) \qquad \mathrm{Ext}_h^i(Q,h) \cong \mathrm{Ext}_h^{i-1}(Q,q_n/h).$$

In the last step of the proof we shall construct h_{n-1}-linear bijections of degree 1

$$(5) \qquad \rho: \mathrm{Hom}_h(Q,q_n)/h) \longrightarrow \mathrm{Hom}_{h_{n-1}}(Q,h_{n-1})$$

for any h-module Q. Applying ρ to the modules Q^i of a free resolution of Q we obtain bijections

$$(6) \qquad \rho_i: \mathrm{Ext}_h^i(Q,q_n/h) \longrightarrow \mathrm{Ext}_{h_{n-1}}^i(Q,h_{n-1}).$$

The combination of (4) and (6) is the statement of the Reduction Lemma.

For the construction of ρ we take a $\Psi \in \mathrm{Hom}_h(Q,q_n/h)$ and expand a representative of ΨQ in a power series

$$\Psi \mathcal{Q} = \frac{1}{y_n} \Phi_0 \mathcal{Q} + \frac{1}{y_n^2} \Phi_1 \mathcal{Q} + \cdots + \chi \mathcal{Q} ,$$

here $\chi \mathcal{Q}$ lies in h. We define

$$\Psi\rho = \Phi_0 .$$

he definition is independent of the representative of the class mod h.
$_0$ is an element of $\mathrm{Hom}_{h_{n-1}}(\mathcal{Q}, h_{n-1})$, and the map ρ is homogeneous of egree 1.

If $\Phi_0 = 0$, we use

$$\Psi y_n^{\lambda} \mathcal{Q} = \frac{1}{y_n} \Phi_{\lambda} \mathcal{Q} + \cdots .$$

o all $\Phi_{\lambda} \mathcal{Q} = 0$, and therefore $\Psi = 0$. This shows that ρ is injective.

Conversely take a $\Phi_0 \in \mathrm{Hom}_{h_{n-1}}(\mathcal{Q}, h_{n-1})$ and define two operators and τ by

$$(\zeta\Phi_0)\mathcal{Q} = \Phi_0(y_n \mathcal{Q}), \qquad \tau\Phi_0 = \frac{1}{y_n - \zeta} \Phi_0 ,$$

here

$$\frac{1}{y_n - \zeta} = \sum_{v=0}^{\infty} \frac{\zeta^v}{y_n^{v+1}} .$$

$\Phi_0 = \Psi$ is an element of $\mathrm{Hom}_h(\mathcal{Q}, q_n/h)$. Indeed, $\tau\Phi_0$ is h_{n-1}-linear. urthermore

$$y_n \tau \Phi_0 \mathcal{Q} = \Phi_0 \mathcal{Q} + \frac{\zeta}{y_n - \zeta} \Phi_0 \mathcal{Q} \equiv \frac{1}{y_n - \zeta} \Phi_0 (y_n \mathcal{Q}) \bmod h .$$

astly we see that

$$(\tau(\Psi\rho) - \Psi)\mathcal{Q} \equiv \frac{1}{y_n^2} \Phi_1' \mathcal{Q} + \cdots \bmod h ,$$

nd this entails

$$\tau(\Psi\rho) = \Psi,$$

as we have seen. Hence ρ is bijective.

In the following we need not return to homological considerations; everything will be based on the reduction lemma.

§8. *Structural properties*

From now on h has again the special meaning

$$h = k_0[y_0,\cdots,y_n].$$

We shall also use $h_{n-1} = k\ [y_0,\cdots,y_{n-1}]$. We wish to connect some properties of finite h-modules M which remain invariant under a linear transformation of the y_ν and under extensions of the constant field k_0. Such are for instance the freeness, the reflexivity, the connections with the derived modules $\mathrm{Ext}_h^i(M,h)$, the ranks $L(\lambda,M)$, etc. So we may always assume in the proofs k_0 to be infinite, while the final statements are also valid for finite k_0. This allows us specially to apply the corollary of Theorem 2 in §5.

Let M be a finite h-module and N a submodule such that there exists a normed polynomial $f \in h$ with $fM \subseteq N$. The quotient module is defined by the short exact sequence of maps of degree 0:

$$(1) \qquad 0 \longrightarrow N \longrightarrow M \longrightarrow M/N \longrightarrow 0$$

(where the second arrow means the injection), from which we derive the long exact sequence of maps of degree 0, applying Theorem 1 of §3 and the reduction lemma:

$$(2) \qquad \begin{array}{l} 0 \longrightarrow M^* \longrightarrow N^* \longrightarrow \eta^{-1}(M/N)^* \longrightarrow \mathrm{Ext}_h^1(M,h) \longrightarrow \cdots \\ \cdots \longrightarrow \mathrm{Ext}_h^i(M,h) \longrightarrow \mathrm{Ext}_h^i(N,h) \longrightarrow \eta^{-1}\mathrm{Ext}_{h_{n-1}}^i(M/N,h_{n-1}) \longrightarrow \cdots \end{array}$$

We use here and in all similar situations an isomorphism η^{-1} of degree

-1 in order that the arrows can always express maps of degree 0. In (2) we have used the abbreviations

$$M^* = \mathrm{Hom}_h(M,h), \quad N^* = \cdots, \quad (M/N)^* = \mathrm{Hom}_{h_{n-1}}(M/N,h_{n-1}),$$

and we have also observed $\mathrm{Hom}_h(M/N,h) = 0$, because M/N is a torsion module due to our assumption. In the following we shall always refer to (2) as to the *fundamental sequence.*

Theorem 1. *If* M *is reflexive and if* $N \subsetneqq M$ *and* $fM \subsetneqq N$ *with a normed* f, *and* N *is free, then* M/N *is a reflexive* h_{n-1}*-module, in particular*

$$M/N \cong \eta^{-1}(N^*/M^*)^*. \tag{3}$$

Proof. If $N \subsetneqq M$, we have $N^* \supsetneqq M^*$, and we can apply (2) as well with N^*, M^* instead of M, N. If N is free, so is N^*, and the beginning of (2) can be expressed as (3). We know that the dual of a module is always reflexive.

In the following we shall study modules with certain regularity properties which will now be defined. We call two finite modules *quasiequal*

$$M \approx N$$

if they coincide in all homogeneous elements of sufficiently large degrees. Quasiequal modules have obviously the same rank polynomials. If M is quasiequal with its reflexive completion:

$$M \approx M^{**},$$

it will be called *quasireflexive.* Lastly we define a quasifree module one for which all $\mathrm{Ext}_h^i(M,h) \approx 0$ $(i > 0)$.

Theorem 2. *Let* M, N *be finite and quasifree modules, and* $N \subsetneqq M$

and $fM \subseteq N$ *with a normed* $f \in h$. *Then* M/N *is a quasifree* h_{n-1}*-module.*

The proof follows immediately from (2).

Proposition. An (as always) graded finite module M *with respect to* $h_0 = k_0[y_0]$ *has a basis* M_i *of homogeneous elements, and* M *is defined by the* M_i *as generators and* $y_0^{\mu_i} M_i = 0$ *for* $i \geq$ *some* i_0 *as relations.*

Furthermore there exists an in general non-homogeneous isomorphism for the torsion-submodule M_0 *of* M:

$$(4) \qquad M_0 \cong \mathrm{Ext}^1_{h_0}(M, h_0),$$

and the following ranks are equal:

$$(5) \qquad L(0, M_0) = L(0, \eta \mathrm{Ext}^1_{h_0}(M, h)),$$

where η *denotes a homogeneous isomorphism of degree* 1.

Proof. Let $M_1, \cdots, M_m$ be a minimal system of homogeneous generators of M. We proceed by induction on m. Such a system contains at least one element of minimal degree; assume M_1 to be such. The quotient module $M' = M/h_0M_1$ has $m-1$ generators and therefore a basis B_i':

$$M' = \bigoplus h_0 B_i' \quad \text{with some} \quad y_0^{\mu_i'} B_i' = 0.$$

We take representatives B_i of the classes B_i'. The latter equations mean

$$y_0^{\mu_i'} B_i = 0 \quad \text{or} \quad = y_0^{\beta_i} M_1 \neq 0.$$

If always the former case holds, a basis of M is $M_1, B_1, \cdots$. Now assume the latter case. Because $\deg M_1 \leq \deg B_i$, we have $\mu_i' \leq \beta_i$ and

$$y_0^{\mu_i'}(B_i - y_0^{\beta_i - \mu_i'} M_1) = 0.$$

Hence B_i can be replaced by $B_i - y^{\beta_i - \mu_i'} M_1$.

For the proof of the second statement we may assume $M = M_0$. With the basis thus constructed we form the free resolution

$$0 \longrightarrow M^1 \xrightarrow[\phi_1]{} M^0 \xrightarrow[\phi]{} M \longrightarrow 0$$

where

$$M^0 = \bigoplus h_0 M_i^0, \quad M^1 = \bigoplus h_0 M_i^1$$

and

$$\phi M_i^0 = M_i, \quad \phi_1 M_i^1 = y_0^{\mu_i} M_i^0 .$$

Let $\deg M_i = \nu_i$, and put

$$\deg (M_i^0) = \deg (M_i) = \nu_i$$

$$\deg (M_i^1) = \nu_i + \mu_i .$$

Now ϕ and ϕ_1 are homogeneous of degree 0. Lastly we form the dual modules

$$M^{0*} = \bigoplus h_0 M_i^{0*}, \quad M^{1*} = \bigoplus h_0 M_i^{1*}$$

with the map

$$\phi_1^* M_i^{0*} = y^{\mu_i} M_i^{1*} .$$

This shows (4).

The rank $L(0,M)$ is equal to the number of i with $-\mu_i < \nu_i \leq 0$, and the rank $L(0,\eta\mathrm{Ext}^1_{h_0}(M,h_0))$ equal to the number of i with $-\mu_i < -\mu_i - \nu_i + 1 \leq 0$. The two numbers are equal. This proves (5).

Theorem 3. *A finite quasifree module is quasireflexive.*

Remark. The converse is not always true as the following counterexample shows. Let M_{n-1} be a reflexive h_{n-1}-module and $h'_n = k_0[y_n]$,

$$h = h_{n-1} \otimes h'_n, \quad M = M_{n-1} \otimes h'_{n-1} .$$

M is readily verified as a reflexive h-module. With a free resolution

M^i_{n-1} of M_{n-1}, $M^i = M_{n-1} \otimes h'_n$ is a free resolution of M. The dual modules are

$$\mathrm{Hom}_h(M^i,h) = \mathrm{Hom}_{h_{n-1}}(M^i_{n-1},h_{n-1}) \otimes h'_n,$$

and the homology groups of the complex are

$$\mathrm{Ext}^i_h(M,h) = \mathrm{Ext}^i_{h_{n-1}}(M_{n-1},h_{n-1}) \otimes h'_n.$$

In general the $\mathrm{Ext}^i_{h_{n-1}}(M_{n-1},h_{n-1})$ do not vanish, and then the $\mathrm{Ext}^i_h(M,h)$ have infinite ranks over k_0.

Proof. For $n = 0$ we saw in the above proposition that $M = \bigoplus hM_i$ with $y_0^{\mu_i} M_i = 0$ for $i \geq$ some i_0. Then $M^* = \bigoplus_{i<i_0} hM^*$ and $M^{**} = \bigoplus_{i<i_0} hM_i$ which is $\approx M$.

Assume the theorem to be proved for n-1 and take a free submodule $N \subseteqq M$ which has the same rank as the torsion-free kernel of M. Then there is an $f \in h$ such that $fM \subseteqq N$, which we may assume to be normed. The beginning of the fundamental sequence (2) gives us because of the assumption

$$0 \longrightarrow M^* \longrightarrow N^* \longrightarrow \eta^{-1}(M/N)^* \longrightarrow \approx 0$$

or

$$N^*/M^* \approx \eta^{-1}(M/N)^* \tag{6}$$

The same sequence with N^*, M^* instead of M, N yields

$$M^{**}/N \cong \eta^{-1}(N^*/M^*)^*. \tag{7}$$

Comparing the dual of (6) with (7) we obtain

$$M^{**}/N \approx (M/N)^{**}. \tag{8}$$

M/N is quasifree due to Theorem 2, and we can apply the induction assumption. By this (8) becomes

$$M^{**}/N \approx M/N. \tag{9}$$

(9) can be interpreted as follows: to every $M_1 \in M^{**}$ of suf-
iciently large degree there exists an $M_2 \in M$ and an $N \in N$ such
hat $M_1 = M_2 + N$. But N is also in M, and so $M_1 \in M$. QED

Theorem 4. *A graded finite and torsion-free module* M *is free if*
nd only if all $\mathrm{Ext}_h^i(M,h) = 0$. (A free graded module is defined as the
irect sum of graded cyclic modules.)

Proof. It is trivial that all Ext^i vanish for a free module.
or the converse we again use induction on n. In the case $n = 0$, such
odules are free as shown in the proposition above.

For the induction we form the quotient $M' = M/y_nM$. It satisfies
he assumption because the $\mathrm{Ext}^i_{h_{n-1}}(M',h_{n-1})$ are squeezed in between
he $\mathrm{Ext}_h^i(M,h)$ and $\mathrm{Ext}_h^i(N,h)$ with $N = y_nM$ in the fundamental se-
uence. Thus $M' = \bigoplus h_{n-1}M_i'$; with homogeneous M_i'. We take
ny homogeneous M_i of the residue classes M_i' mod y_nM and contend that
hey form a basis of M. Indeed an $M \in M$ can be represented as

$$M = \sum m_{i0}M_i + y_nM^{(1)}$$

ith $m_{i0} \in h_{n-1}$ and $M^{(1)} \in M$. The same is possible for $M^{(1)}$:

$$M^{(1)} = \sum m_{i1}M_i + y_nM^{(2)},$$

nd so on:

$$M = \sum (m_{i0} + y_n m_{i1} + \cdots + y_n^r m_{ir})M_i + y_n^{r+1}M^{(r+1)}$$

or $r = 1,2,\cdots$. The degrees of the $M^{(r)}$ decrease. Since M is finite,
he $M^{(r+1)} = 0$ from some r_0, which proves the theorem.

9. *The theorem of Riemann-Roch and the duality theorem*

Here again the remark at the beginning of § 8 applies that we may
ithout loss of generality assume the constant field k_0 as infinite in
he proofs, while the statements are also valid for finite k_0.

The problem of Riemann-Roch is to obtain information on the rank

$$\dim M \; = \; L(0,M)$$

for a finite module M. This rank is connected with that of the *canonically conjugated module*.

$$(1) \qquad M_* \; = \; y_0^{n+1} M^* \; = \; y_0^{n+1} \operatorname{Hom}_h(M,h).$$

The distinction of the variable y_0 seems incomprehensible. But because $\dim M$ is an invariant of the class of M, i.e., $\dim M = \dim aM$ for a homogeneous a of degree 0, $\dim M_*$ is independent of the variable y_0 used in its definition.

In this connection we may mention the following

Lemma. If M is torsion-free and if h is a homogeneous polynomial, there exists a homogeneous isomorphism $\eta^{-\deg h}$ of degree $-\deg h$ such that

$$\operatorname{Ext}_h^i(hM,h) \;\tilde{=}\; \eta^{-\deg h} \operatorname{Ext}_h^i(M,h)$$

where $\tilde{=}$ means an isomorphism of degree 0.

Proof. For $i = 0$ we have $\operatorname{Hom}_h(hM,h) = h^{-1}\operatorname{Hom}_h(M,h)$. The same fact for the modules of a free resolution of M implies the contention for $i > 1$.

Theorem of Riemann-Roch. For a finite h-module M the following equation holds:

$$(2) \qquad \dim M + (-1)^n\{\dim M_* - L(0,\eta^{n+1} \operatorname{Ext}_h^1(M,h) + - \cdots\} \; = \; \gamma(M).$$

The alternating sum terminates after finitely many terms. Indeed $\operatorname{Ext}_h^i(M,h) = 0$ for $i > n+1$, and even for $i > n$ if M is torsion free, and for $i > n-1$ if M is reflexive. (η^{n+1} means an isomorphism of degree $n+1$).

Proof. We begin with verifying (2) for a free $M = hM$ where M

has degree μ. Then the rank function is

(3)
$$L(\lambda, hM) = \begin{cases} \binom{\lambda-\mu+n}{n} & \text{for } \lambda \geq \mu, \\ 0 & \text{for } \lambda < \mu . \end{cases}$$

Inserting $\lambda = 0$ into the formal polynomial on the right we get

(4)
$$\gamma(hM) = \frac{(1-\mu)\cdots(n-\mu)}{n!} .$$

Furthermore $\mathrm{Hom}_h(hM,h) = hM^*$ with an M^* of degree $-\mu$, and $(hM)_* = hy_0^{n+1}M^*$. This yields

$$L(\lambda, (hM)_*) = \begin{cases} \binom{\lambda+\mu-1}{n} & \text{for } \lambda \geq -\mu+n+1 \\ 0 & \text{for } \lambda < -\mu+n+1 . \end{cases}$$

But the right side vanishes also for $\lambda = 1-\mu, \cdots, n-\mu$. So

(5)
$$L(\lambda, (hM)_*) = \begin{cases} \binom{\lambda+\mu-1}{n} & \text{for } \lambda > -\mu \\ 0 & \text{for } \lambda \leq -\mu . \end{cases}$$

The $\mathrm{Ext}_h^i(hM,h) = 0$ because hM is free. Now we take from (3) and (5)

$$\dim hM = \begin{cases} \binom{n-\mu}{n} = \frac{(n-\mu)\cdots(1-\mu)}{n!} & \text{for } 0 \geq \mu \\ 0 & \text{for } 0 < \mu , \end{cases}$$

$$\dim (hM)_* = \begin{cases} 0 & \text{for } 0 \leq -\mu, \\ \binom{\mu-1}{n} = \dfrac{(\mu-1)\cdots(\mu-n)}{n!} & \text{for } 0 > -\mu. \end{cases}$$

Comparison with (4) proves the theorem in this case.

Because the functions $\gamma(M)$, $\dim M$, $\dim M_*$ are linear in direct sums (for the Ext^i see §4, Theorem 2) the Riemann-Roch theorem is proved for free modules.

In the next step we verify (2) in the case $n = 0$, applying the proposition of §8. Let $M = \bigoplus hM_i$ with the M_i being free for $i \leq i_0$ and torsion elements for $i > i_0$. Then $\gamma(M) = i_0$. Furthermore

$$\dim M = (\text{number of } i \leq i_0 \text{ with } \deg M_i \leq 0) + L(0,M_0)$$

where M_0 is the torsion-submodule of M. The dimension of M_* is

$$\dim M_* = (\text{number of } i \leq i_0 \text{ with } 1 - \deg M_i \leq 0).$$

$1 - \deg M_i \leq 0$ can be expressed as $\deg M_i > 0$. Lastly we apply equation (5) in §8. The collection of all terms yields (2).

Now we proceed by induction on n, beginning with $n = 0$, where it is easily checked that $\mathrm{Ext}_h^2(M,h) = 0$ for all M and $\mathrm{Ext}_h^1(M,h) = 0$ if M is torsion-free. For the proof of $\mathrm{Ext}_h^n(M,h) = 0$ for a reflexive M, induction begins at $n = 1$. Here we take a free module $N \supseteq M$ with some normed $f \in h$ such that $fN \subseteq M$. Now the fundamental sequence (2) is

$$0 \longrightarrow \cdots \longrightarrow \mathrm{Ext}_h^1(N,h) \longrightarrow \mathrm{Ext}_h^1(M,h) \longrightarrow \eta^{-1}\mathrm{Ext}_{h_0}^1(N/M,h_0) = 0,$$

whence $\mathrm{Ext}_h^1(M,h) = 0$.

For the induction proper we take a free $N \subseteq M$ which has the same rank as the torsion-free kernel of M. Therefore there exists an $f \in h$ such that $fM \subseteq N$, which we may assume to be normed. With this the fundamental sequence (2) of §8 can be formed. Because all $\mathrm{Ext}_h^i(N,h) = 0$ for $i > 0$, and all $\mathrm{Ext}_{h_{n-1}}^i(M/N,h_{n-1}) = 0$ for $i > n$, we see that the $\mathrm{Ext}_h(M,h)$ vanish for $i > n+1$. If M is torsion-free, M/N is torsion-free due to Theorem 2 of §5, and the vanishing of the Ext^i begins already at $i = n+1$. If M is even reflexive, also M/N is reflexive by Theorem 1 of §8, which has $\mathrm{Ext}_h^n(M,h) = 0$ as a consequence.

The alternating sum of the ranks $L(-n-1,\mathrm{Hom}_h(M,N))$ etc., in the fundamental sequence vanishes, as found in §1, Proposition 3. This can be expressed as

$$(6)\qquad \begin{aligned}(\dim M_* - L(0,\eta^{n+1}\,\mathrm{Ext}_h^1(M,N) + - \cdots) - \dim N_* &= \\ = \dim (M/N)_* - L(0,\eta^n\,\mathrm{Ext}_{h_{n-1}}^1&(M/N,h_{n-1}) + - \cdots.\end{aligned}$$

It is at this point that the concept of the canonically conjugated module is used. For the dual modules the notation would be less appropriate. Substracting the Riemann-Roch formula for N instead of M from (2) and observing §6, (6), we obtain (6). This concludes the proof.

Theorem of Duality. Let $n > 1$. *For a finite, reflexive, and quasifree h-module M the following equations hold* $(i = 1,\cdots,n-1)$:

$$(7)\qquad L(\lambda,\mathrm{Ext}_h^i(M,h)) = L(-\lambda-2n-2,\mathrm{Ext}_h^{n-i}(M_*,h)).$$

Corollary. If M has these properties, also M_* *is quasifree.*

This is an obvious consequence of (7).

Remarks. Because of the lemma at the beginning of §9 we have

$$L(\lambda,\mathrm{Ext}_h^i(M,h)) = L(-n-1,\mathrm{Ext}_h^i(y_0^{n+1+\lambda}M)),$$

$$L(-\lambda-2n-2,\mathrm{Ext}_h^i(M_*,h)) = L(-n-1,\mathrm{Ext}_h^i(y_0^{-n-1-\lambda}M_*)).$$

Thus (7) is equivalent with the special case $\lambda = -n-1$ with $y_0^{-n-1+\lambda}M$ instead of M.

Because of (2), (7) implies

(8) $$\gamma(M) = (-1)^n\gamma(M_*).$$

If M is the ideal of multiples of a divisor in a regular variety, (7) can be interpreted in the language of sheaves. In this form it is equivalent with J.P. Serre's Duality Theorem (see Faisceaux algébriques coherents, Annals of Math. 61(1956), 197-278; No. 75).

We shall see simultaneously

Theorem 1. *If* $n = 2$, *every finite and reflexive h-module is quasifree.*

For $n > 2$ this is not true as shown by the counterexample following Theorem 3 in §8.

In the proofs we shall need the

Proposition. For a finite and torsion-free module M *we have*

$$\gamma(y_0^{-\lambda}M) = H(\lambda,M).$$

Proof. At first let M be free. It suffices to prove the statement for a one-dimensional $M = Mh$ where $\deg M = \mu$. The rank polynomial is

$$H(\lambda,M) = \frac{(\lambda-\mu+n)\cdots(\lambda-\mu+1)}{n!},$$

and especially

$$\gamma(M) = \frac{(-\mu+n)\cdots(-\mu+1)}{n!}.$$

Now the contention is evident.

For more general M we take a free submodule $N \subseteq M$ of equal rank as M and use §6, (1) and (6):

$$L(\lambda,M) - L(\lambda,N) = L(\lambda,M/N),$$

$$\gamma(y_0^{-\lambda}M) - \gamma(y_0^{-\lambda}N) = \gamma(y_0^{-\lambda}(M/N))$$

and the claimed formula for $n-1$ instead of n.

Proof of the Duality Theorem and Theorem 1. As remarked before, we need only prove (7) for $\lambda = -n-1$. If $n = 2$ we do this without the assumption that M is quasifree. Applying (7) with variable λ we see that $L(\lambda,\mathrm{Ext}_h^1(M,h))$ and $L(\lambda,\mathrm{Ext}_h^1(M_*,h))$ can be $\neq 0$ only for finitely many λ. This is $\mathrm{Ext}_h^1(M,h) \approx 0$ or Theorem 1.

Let $n = 2$ and take a free submodule N. Without loss of generality we may assume $fM \subseteq N$ with a normed $f \in h$. Now we form the fundamental sequence (2) in §8.

$$(9) \qquad 0 \longrightarrow M^* \longrightarrow N^* \longrightarrow \eta^{-1}(M/N)^* \longrightarrow \mathrm{Ext}_h^1(M,h) \longrightarrow 0.$$

In §8 we saw that M/N is reflexive and isomorphic with $\eta^{-1}(N^*/M^*)^*$. So $\eta^{-1}(M/N)^*$ is isomorphic with the reflexive completion of N^*/M^*. Now we form the rank polynomials, using §5, Proposition 2:

$$H(\lambda,N^*/M^*) = N'\lambda - G(N^*/M^*),$$

$$H(\lambda,\eta^{-1}(M/N)^*) \quad = \quad N'\lambda - G(\eta^{-1}(M/N)^*).$$

That proposition also tells us that a module and its reflexive completion have the same linear degrees. Hence both rank polynomials are equal, and then (9) implies $\mathrm{Ext}^1_h(M,h) \approx 0$.

Thus prepared we apply the Riemann-Roch formula for $y_0^{-\lambda}M$ and $y_0^{\lambda}M_*$:

$$\dim y_0^{-\lambda}M + \dim y_0^{\lambda}M_* - L(0,\eta^{\lambda+3}\mathrm{Ext}^1_h(M,h)) \quad = \quad \gamma(y^{-\lambda}M),$$

$$\dim y_0^{-\lambda}M + \dim y_0^{\lambda}M_* - L(0,\eta^{-\lambda+3}\mathrm{Ext}^1_h(M_*,h)) \quad = \quad \gamma(y_0^{\lambda}M_*).$$

Since $\mathrm{Ext}^1_h(M,h) \approx \mathrm{Ext}^1_h(M_*,h) \approx 0$, the terms containing the Ext vanish for large positive or negative λ. The remaining terms on the left are symmetric. Hence the right terms are also symmetric, because they are polynomials, due to the above proposition. Now eventually the terms containing the Ext exhibit the same symmetry, namely (7).

In the following induction proof we assume $n > 2$. Also we assume M_* to be quasifree instead of M which means, in other words, that we replace M by M_*. We take a free $N \subseteq M$ such that $fM \subseteq N$ with a normed f, and form

$$Q \quad = \quad N_*/M_*.$$

This is a torsion-free h_{n-1}-module which is quasifree by Theorem 2 of §8. For a later application we compare it with its reflexive completion and put

$$R \quad = \quad Q^{**}/Q.$$

Because Q is quasireflexive (§8, Theorem 3), $R \approx 0$. We consider the exact sequence attached to the definition of R:

$$0 \longrightarrow \cdots \longrightarrow \eta^{-1}\,\mathrm{Ext}^{i}_{h_{n-2}}(R,h_{n-2}) \longrightarrow \mathrm{Ext}^{i+1}_{h_{n-1}}(Q^{**},h_{n-1}) \longrightarrow$$

$$\longrightarrow \mathrm{Ext}^{i+1}_{h_{n-1}}(Q,h_{n-1}) \longrightarrow \eta^{-1}\,\mathrm{Ext}^{i+1}_{h_{n-2}}(R,h_{n-2}) \longrightarrow \cdots$$

In $\mathrm{Ext}^{i}_{h_{n-2}}(R,h_{n-2})$ we can apply the reduction theorem i times and get

$$\mathrm{Ext}^{i}_{h_{n-2}}(R,h_{n-2}) \cong \eta^{-i}\,\mathrm{Ext}^{0}_{h_{n-i-2}}(R,h_{n-i-2}) = 0,$$

because $R \approx 0$. This is possible for $i = 1,\cdots,n-2$. Thus we obtain

$$(10)\quad \mathrm{Ext}^{i}_{h_{n-1}}(Q^{**},h_{n-1}) \cong \mathrm{Ext}^{i}_{h_{n-1}}(Q,h_{n-1}), \quad i = 1,\cdots,n-2.$$

Now we use the two exact sequences of maps of degree 0:

$$(11)\quad \begin{aligned} 0 \longrightarrow M^{*} \longrightarrow \cdots \longrightarrow \mathrm{Ext}^{i}_{h}(N,h) = 0 \longrightarrow \eta^{-1}\,\mathrm{Ext}^{i}_{h_{n-1}}(M/N,h_{n-1}) \longrightarrow \\ \longrightarrow \mathrm{Ext}^{i+1}_{h}(M,h) \longrightarrow \mathrm{Ext}^{i+1}_{h}(N,h) = 0 \longrightarrow \cdots \end{aligned}$$

and

$$(12)\quad \begin{aligned} 0 \longrightarrow N^{**} = N \longrightarrow \cdots \longrightarrow \mathrm{Ext}^{i}_{h}(N^{*},h) = 0 \longrightarrow \mathrm{Ext}^{i}_{h}(M^{*},h) \longrightarrow \\ \longrightarrow \eta^{-1}\,\mathrm{Ext}^{i}_{h_{n-1}}(N^{*}/M^{*},h_{n-1}) \longrightarrow \mathrm{Ext}^{i+1}_{h}(N^{*},h) = 0 \longrightarrow \cdots. \end{aligned}$$

Replacing M^{*} by M_{*} we get from (12)

$$(13)\quad \mathrm{Ext}^{i}_{h}(M_{*},h) \cong \eta^{-1}\,\mathrm{Ext}^{i}_{h_{n-1}}(Q,h_{n-1}).$$

In (11) we have because of (the beginning of) (12) $M/N \cong \eta^{-1}(N^{*}/M^{*})^{*}$ or $M/N \cong (N_{*}/M_{*})_{*}$. With this we get from (11)

$$(14)\quad \eta^{-1}\,\mathrm{Ext}^{i-1}_{h_{n-1}}(Q_{*},h_{n-1}) \cong \mathrm{Ext}^{i}_{h}(M,h), \quad i > 1.$$

This implies, for $i > 1$ (here we use $n > 2$):

$$L(\lambda,\mathrm{Ext}^{i}_{h}(M,h)) = L(\lambda+1,\mathrm{Ext}^{i-1}_{h_{n-1}}(Q_{*},h_{n-1})).$$

The right side is by induction

$$= L(-\lambda-2n-1,\mathrm{Ext}^{n-i}_{h_{n-1}}(Q_{**},h_{n-1})),$$

with $Q_{**} = Q^{**}$, and because of (10)

$$= L(-\lambda-2n-1,\mathrm{Ext}^{n-i}_{h_{n-1}}(Q,h_{n-1})),$$

finally due to (13)

$$= L(-\lambda-2n-2,\mathrm{Ext}^{n-i}_{h}(M_{*},h)).$$

So (7) is proved for $i = 2,\cdots,n-2$.

In order to obtain the formula for $i = 1$, we apply equation (5) of §8 in the case $M_0 = R$:

$$L(0,R) = L(0,\eta\mathrm{Ext}^1_{h_0}(R,h_0)). \tag{15}$$

Thus prepared we return to (11) and (12). The first 4 terms of the sequence (11) yield (after applying η $(n+1)$-times)

$$(M/N)_*/(N_*/M_*) \cong \eta^{n+1}\ \mathrm{Ext}^1_h(M,h)$$

or (since $Q_* = M/N$ or $Q_{**} = (M/N)_*$ as we have seen)

$$R = Q_{**}/Q = \eta^{n+1}\ \mathrm{Ext}^1_h(M,h). \tag{16}$$

This isomorphism is homogeneous of degree 0. Because R is a finite h_0-torsion module and $\mathrm{Ext}^{n-1}_{h_{n-1}}(Q_{**},h_{n-1}) = 0$, the arguments leading to (10) yield a homogeneous isomorphism of degree 0

$$\mathrm{Ext}^{n-1}_{h_{n-1}}(Q,h_{n-1}) = \eta^{1-n}\ \mathrm{Ext}^1_{h_0}(R,h_0). \tag{17}$$

Combining (13) for $i = n-1$ and (15)-(17) we get finally

$$L(0,\eta^{n+1}\mathrm{Ext}^{n-1}_h(M_*,h)) = L(0,\eta^n\mathrm{Ext}^{n-1}_{h_{n-1}}(Q,h_{n-1})) =$$

$$= L(0,\eta\mathrm{Ext}^1_{h_0}(R,h_0)) = L(0,R) = L(0,\eta^{n+1}\mathrm{Ext}^1_h(M,h)).$$

This is (7) with $\lambda = -n-1$; and this suffices as we remarked on page 44.

CHAPTER II

GRADED RINGS AND IDEALS

§10. *Introduction, Divisors*

We will consider integral domains J with the following properties:

1) *J is graded, and the elements of degree 0 form a field k_0, the constant field. k_0 will always be assumed as infinite.*

2) *J can be generated over k_0 by finitely many elements.*

Instead of the *degree* of the elements of J we use the word *weight*. The reason will become clear soon.

As in Chapter I we make the convention that only sums of elements of equal weights are allowed. If J is integrally closed in its quotient field, J will be called *normal*. In general we will assume J as normal, unless otherwise stated.

Elements of such domains are called *algebraic forms*. The quotients of forms of equal weights are *algebraic functions*. They form the *quotient field* K of J. Examples of such domains are given by the automorphic forms of various kinds. That they have the property 2) is a deep theorem which has been proved for Siegel modular forms by H. Cartan in 1957, for a large class of automorphic forms by W.L. Baily and A. Borel in 1966, and under restricted conditions in an easier way by the author in 1969. We shall take this fact for granted.

Let $u_0,\cdots,u_m$ be a system of generators of J and $\nu_0,\cdots,\nu_m$ their weights. Put $\nu = \nu_0\cdots\nu_m$ and

$$U_i = u_i^{\nu/\nu_i}.$$

These elements have equal weights. Consider the linear forms in the U_i

$$y_i = \sum_j \kappa_{ij} U_j, \qquad (i = 0,\cdots,n \leq m).$$

The normalization theorem states that such constants $\kappa_{ij} \in k_0$ can be found that all U_i depend integrally on the y_i, and that no algebraic equations hold between the y_i (see for example O. Zariski and P. Samuel, Commutative Algebra, Vol. II). The number n determined in this way is the *dimension* of J. Now not only the U_i but also the u_i and hence all elements of J depend integrally on the y_i or, in other words, on the rational subring

$$h = k_0[y_0,\cdots,y_n]. \tag{1}$$

The subring h and the equations determining J as an algebraic extension of h is called a *(projective) model* of J. The common weight of the y_i is the *weight of the model*; it will always be denoted by h. An element of J whose weight is a multiple of h, say λh, is an algebraic form of the variables y_i of *degree* λ. This is especially the case for polynomials in the y_i.

As a model of J is not uniquely determined, we are interested in concepts which are invariant, in other words which do not depend on the model. The classical example of such a concept is that of *valuation*. We write it in the multiplicative way:

$$|ab|_p = |a|_p|b|_p, \qquad |a+b|_p \leq \text{Max}\,(|a|_p,|b|_p).$$

We are only interested in valuations of the following sort (without mentioning this again):

1) *they are trivial on* k_0;

2) *the quotient ring*

$$J/\{a \in J, |a|_p < 1\}$$

has dimension n-1.

uch valuations are usually called *valuations of rank one.*

Theorem 1. *In h, all valuations coincide with the p-adic valua-
ions attached to prime polynomials $p \in h$. In J, they coincide with
he extensions of the p-adic valuations.*

Proof. Let $p \in h$ with $|p| < 1$. If p is not a prime polynomial,
factor of p has also value < 1. Because h is a UFD, we can find a
rime p with $|p| < 1$.

There can be no other prime q with $|q| < 1$. Otherwise there
ould be $u, v \in h$ such that

$$up + vq = r \in h_{n-1},$$

nd also $|r| < 1$. This would entail that the residue ring of J with
espect to the valuation has dimension $< n$-1.

The rest of the theorem is evident.

All valuations are *discrete*, and to each valuation p there corre-
pond elements p_p whose values $|p_p|_p$ are maximal < 1. They are called
he *prime elements* attached to p. As such they are uniquely determined
y p up to a unit factor.

We shall use p as the symbol for a valuation of h. The extensions
f p in J will be written p. That p is an extension of p will be ex-
ressed thus:

$$p|p \quad (p \text{ divides } p).$$

To the valuations p *prime divisors* are attached which are also
bbreviated by p. A *divisor* is a formal product

$$m = \prod p^{\nu_p(m)} \qquad 2)$$

where only finitely many $\nu_p(m) \neq 0$. The multiplication of divisors is defined in the obvious way. The divisors form an infinite abelian group.

To elements $a \in J$ the divisor

$$(a) = \prod p^{\nu_p(a)} \quad \text{with} \quad |a|_p = |p_p|^{\nu_p(a)}$$

is attached where p_p is a prime element of p. The equation

$$(ab) = (a)(b)$$

is evident.

Two divisors m_1 and m_2 are called *equivalent*:

$$m_1 \sim m_2$$

if their quotient is a divisor attached to an element of K. The *divisor classes* also form an abelian group.

The prime polynomial p of h, as a divisor, has a prime decomposition

$$(p) = p_1^{e_1} \cdots p_r^{e_r}$$

where the exponents e are called the *ramification indices*.

Along with K we also use the rational subfield $k = k_0(x_1, \cdots, x_n)$, $x_i = \frac{y_i}{y_0}$ and the graded extensions

$$\bar{k} = \bigoplus_{\nu=-\infty}^{+\infty} k y_0^{\nu}, \quad \bar{K} = \{\tfrac{a}{b}:\ a, b \in J,\ b \text{ homogeneous}\}.$$

Let m be a divisor and (2) its prime decomposition. Then

$$M(m) = \{a \in \bar{K}: |a|_p \leq |p_p|^{\nu_p(m)}\}$$

is called *the ideal of multiples of* m.

These $M(m)$ are obviously the intersections

$$(3) \qquad M(m) = \bigcap_p M(m)_p,$$

where

$$(4) \qquad M(m)_p = \bigcap_{\mathfrak{p}|p} \{a \in \bar{K}: |a|_{\mathfrak{p}} \leq |p_{\mathfrak{p}}|^{\nu_{\mathfrak{p}}(m)}\}.$$

Conversely we claim

$$(5) \qquad M(m)_p = h_p M(m).$$

The inclusion $h_p M(m) \subseteq M(m)_p$ is clear from the definition. Let $a \in M(m)_p$. as defined by (4). If $a \notin M(m)$, there are finitely many prime divisors q for which $a \notin M(m)_q$. These q do not divide p. Thus there is a polynomial $d \in h$, prime to p, such that $da \in M(m)_q$. So $da \in M(m)$. But $d^{-1} \in h_p$, and $a \in h_p M(m)$. Now we can apply Theorem 1 of §5 and find that the $M(m)$ are reflexive h-modules.

Obviously they are also J-ideals. If J is normal, we have

$$J = M(1),$$

the ideal of multiples of the unit divisor.

Theorem 2. *If J is normal, J and all ideals of multiples of divisors are finite and reflexive h-modules. Conversely, such ideals are ideals of multiples of divisors.*

Proof. That J is finite stems from the assumption that J is finitely generated. It is then usually called *Noetherian*. All ideals in a Noetherian ring are known to be finite modules.

This applies to *integral ideals*, that means ideals $M \subseteq J$ and *integral divisors*: $\nu_{\mathfrak{p}}(m) \geq 0$ for all $\mathfrak{p}$. For non-integral divisors there always exist polynomials $f \in h$ such that $fM(m) \in J$, and so also these ideals are finite h-modules.

Conversely a finite and reflexive J-ideal M is an intersection (3) of modules M_p by §5, Theorem 1. The M_p must be also J-ideals. Then they are automatically J_p-ideals with

$$J_p = \{a \in \overline{K}: |a|_p \leq 1 \text{ for all } p|p\}.$$

Now we take from elementary commutative algebra that

$$J_p = \bigcap_{p|p} J_p, \qquad M_p = \bigcap_{p|p} M_p.$$

J_p is the so-called *valuation ring* of p. M_p is an J_p-ideal, and all J_p-ideals are principal ideals of the form

$$M_p = J_p p_p^{\mu_p}.$$

This completes the proof.

In the case $J = h$ Theorem 2 states that all reflexive ideals are principal ideals $M = hp_1^{\mu_1}\cdots$. We may refer to this fact by saying that h is a *reflexive principal ideal domain* or RPID. At a later occasion we shall have to use a little more, namely,

Theorem 3. *Let* $u_0,\cdots,u_n$ *be independent variables to which weights* $\nu_0,\cdots,\nu_n > 0$ *are assigned which are not divisible by the characteristic of* k_0. *The graded ring of weighted polynomials in the* u_i *is a* RPID.

Proof. By a weighted polynomial we mean an expression

$$\sum \kappa_\rho u_0^{\rho_0}\cdots u_n^{\rho_n}$$

with $\rho_0\nu_0 + \cdots + \rho_n\nu_n$ = the same for all summands. We adjoin the roots $y_i = u_i^{1/\nu_i}$ to $\overline{K}$ and the ν_i-th root of unity to k_0. Now $k_0[y_i]$ is a RPID, and so a reflexive J-ideal M is a principal ideal generated by a homogeneous polynomial f:

$$M = k_0[y_i]\cdot f.$$

Applying all automorphisms of $\overline{K}$ which multiply the y_i by certain roots of unity $e^{2\pi i m/\nu_i}$, we see that f is at most transformed into f times a root of unity. It follows that f is a weighted polynomial g in the u_i, times a monomial $y_0^{r_0} \cdots y_n^{r_n}$. Then M is divisible by this monomial. But because M is made up of polynomials in the u_i, also this monomial is one in the u_i.

So we have at least

$$M = k_0[u_i]f$$

but with an f whose coefficients may lie in a larger field. Now again we apply the automorphisms of the Galois group of that field extension and find that f can at most be multiplied by a constant under such an automorphism. Then there is an element α in the larger field which behaves in the same way, and $\frac{1}{\alpha} f$ is a weighted polynomial in the original ring. This completes the proof.

In order to apply the theory of modules developed in Chapter I we can split up J in the following way. Let h be the weight of the model and J_r the set of all forms in J whose weights are $\equiv r \bmod h$. Then

$$J = \bigoplus_{r=0}^{h-1} J_r. \tag{6}$$

Here

$$J_0 = H \tag{7}$$

is again a graded integral domain, and the J_r are H-ideals. Similarly every J-ideal is split up in the same way

$$M = \bigoplus_{r=0}^{h-1} M_r. \tag{8}$$

It is advantageous to treat H and the H-ideals M_0 first. It is in general an easy task to translate properties proved for H-ideals to J-ideals.

§11. *Differentials and the theorem of Riemann-Roch*

In order to avoid difficulties or at least the necessity of special considerations we make the following assumptions:

1) *The constant field k_0 is perfect.*

2) *The characteristic of k_0 is either 0 or greater than the degree of the field extension*

$$N = [K:k]$$

This includes that K/k is a separable field extension. In general algebraic geometry also the inseparable case is treated. But we are ultimately only interested in the case when the characteristic is 0.

Our next aim is to find more invariants of J or of the quotient field K.

In §10 we mostly dealt with J or the division ring $\bar{K}$. In §11 we concentrate more on the quotient field K. We recall briefly the concept of the valuation ring of a valuation p:

$$K = \{a \in K: |a|_p \leq 1\}$$

and the prime ideal

$$P = \{a \in K: |a|_p < 1\}.$$

The quotient ring is often abbreviated as

$$K_p/P_p = K/p.$$

It is always a field. As we only consider valuations of rank one, K/p is a field of dimension n-1, in other words, it contains exactly n-1 algebraically independent elements.

Proposition.

a) *Let p be a prime divisor and z_1 a prime element with respect to p. Then we can find further elements $z_2,\cdots,z_n$ in K with the property that the quotient field K/p is a separable extension of*

$k_0(\bar{z}_2,\cdots,\bar{z}_n)$ *(where* $\bar{z}_i$ *are the residues of the* z_i mod P_p*).*

) *K is a separable extension of* $k_0(z_1,\cdots,z_n)$.

) *If* $z'_1,\cdots,z'_n$ *are other elements of the properties described under* a), *the Jacobian* $J = \left|\frac{\partial z'_i}{\partial z_j}\right|$ *is a p-adic unit.*

We will call the z_i a system of *normal coordinates of K in p.*

Proof. Because we assume k_0 as perfect any finitely generated ield over k_0 can be generated in a separable way (cf. Zariski and amuel, Commutative Algebra, Vol. I). We apply this to the residue ield K/p and find in it n-1 algebraically independent elements $\bar{z}_2$, $\cdots,\bar{z}_n$ such that K/p is separable over $k_0(\bar{z}_2,\cdots,\bar{z}_n)$.

Take any $z_i \in K_p$ whose residues mod P_p are the $\bar{z}_i$. Suppose, ontrary to our contention, that K were not separable over $k_0(z_1,\cdots,$ $_n) = k$. Then there exist two fields $K_1 \subset K_2$ between k and K with $K_2:K_1] = p$ = characteristic and K_2/K_1 inseparable. Call p_1 and p_2 the rime divisors of K_1 and K_2 which are divisible by p. Because K_2/K_1 is nseparable, we have either $p_1 = p_2^p$ or $p_1 = p_2$. The former is impossible because k and then K_1 contains already a prime element of p, amely z_1. The latter is, because then K_2/p_2 would be an inseparable xtension of K_1/p_1, contrary to our construction.

For the proof of the last assertion of our proposition we have to efer to the differential calculus for algebraic functions in one variable (for instance M. Eichler, Introduction to the theory of algebraic umbers and functions, Chapter III, §4). The partial derivatives $\frac{\partial u}{\partial z_1}$ re formed while keeping $z_2,\cdots$ constant. That means K has to be onsidered as an algebraic function field in one variable over the onstant field $k_0(z_2,\cdots,z_n)$. The classical theory attaches to a diferential du a divisor (du), and the equation

$$\left(\frac{\partial u}{\partial z_1}\right) = \frac{(du)}{(dz_1)}$$

holds. As z_1 is assumed a prime element of p, the divisor (dz_1) is prime to p, and $\left(\frac{\partial u}{\partial z_1}\right)$ is integral at p, if u is.

Now we consider $\frac{\partial u}{\partial z_2}$ for a $u \in K_p$, and we assume that u satisfies a separable equation

$$(1) \qquad f(u;z_2,\cdots,z_n) \equiv u^m + c_1(z)u^{m-1} + \cdots \equiv 0 \bmod P_p$$

in the residue field. u satisfies also a separable equation

$$F(u;z_1,\cdots,z_n) = u^M + C_1(z)u^{M-1} + \cdots = 0.$$

We may assume both equations as irreducible. Then the second, taken mod P_p, will be

$$F \equiv f\cdot g \bmod P_p$$

with another polynomial g. Now we have

$$\frac{\partial u}{\partial z_2} = -\frac{F_{z_2}}{F_u} \equiv -\frac{f_{z_2}}{f_u} \bmod P_p.$$

The right quotient is in K_p because we assumed (1) as separable.

If (1) is inseparable, that means if (1) contains only powers u^μ with μ divisible by the characteristic, then either u is inseparable over k, but this contradicts the construction of K, or the $c_i(z)$ in (1) are also p-th powers, and also f is a p-th power of a polynomial f_1, and we may replace f by f_1.

So in any case $\frac{\partial u}{\partial z_2}$ is integral in p. And then also the Jacobian $J = \frac{\partial z_i'}{\partial z_j}$ is integral in p. Because we may exchange the z_i and z_i', J is even a p-adic unit, as contended.

The proposition serves us to attach a divisor to a differential form. The procedure is the generalization of that known in the classical theory of algebraic functions in one variable. A *differential* is a formal expression

$$\omega = a\,dz_1 \wedge \cdots \wedge dz_n$$

where $a, z_1, \cdots, z_n \in K$, and K is separable over $k_0(z_1, \cdots, z_n)$. Two such expressions are equal:

$$\omega' = a' dz_1' \wedge \cdots \wedge dz_n' = \omega$$

if

$$a' \left| \frac{\partial z_i'}{\partial z_j} \right| = a.$$

(The symbol $\wedge$ expresses a formal product, and also sums $\sum a_\nu dz_{\nu_1} \wedge \cdots \wedge dz_{\nu_i}$ are considered in Exterior Algebra. But we shall not use all this.)

Now we attach a divisor to a differential:

$$(\omega) = \prod p^{\nu_p(\omega)}$$

where $\nu_p(\omega)$ is the power of p dividing a if $z_1, \cdots, z_n$ are chosen according to the above proposition. This proposition, together with the definition of equality of divisors, shows that the divisor is uniquely determined.

Theorem 1. *All divisors of differentials are equivalent.* Their class is called the *divisor class* or *canonical class* of the field.

This is almost obvious.

Theorem 2. *In the case of the* x_i *having the same meaning as in* §10, *the ideal of multiples of* $(dx_1 \wedge \cdots \wedge dx_n)^{-1}$ *is*

$$y_0^{n+1} H^* = H_* = M((dx_1 \wedge \cdots \wedge dx_n)^{-1}).$$

Proof. It is based on Dedekind's theorem on the different: if

$$(1) \qquad (p) = p_1^{e_1} \cdots p_r^{e_r} \qquad (e_\rho = e_{p_\rho})$$

is the decomposition of a prime divisor (p) of k, the ramification indices e_i are 1 up to finitely many exceptions, and

$$(2) \qquad H^* = M(\prod p^{1-e_p}).$$

The proof of Theorem 2 has to compare the contributions of each p to $(dx_1\wedge\cdots)$ and $\prod p^{1-e_p}$.

First we treat the case when p does not divide y_0. Let q be the polynomial in the x_i of smallest degree which is divisible by p. Obviously it is a prime polynomial. Not all x_i occur in q in powers which are divisible by the characteristic p. Otherwise q would be a p-th power because k_0 is perfect, which contradicts the minimality of the degree of q. So not all $\frac{\partial q}{\partial x_i} = 0$, say $\frac{\partial q}{\partial x_1} \neq 0$. Then also $\frac{\partial q}{\partial x_1} \not\equiv 0 \bmod p$. Otherwise q and $\frac{\partial q}{\partial x_1}$ are two algebraically independent elements, and there can only exist $n-2$ other algebraically independent ones. It would follow that the quotient field K/p contains only $\leq n-2$ algebraically independent elements, contrary to our assumption in §10. The residue field K/p is an extension of the residue field k/q of degree $\leq [K:k]$. Because $\frac{\partial q}{\partial x_1} \not\equiv 0 \bmod p$, the residue classes of $x_2,\cdots,x_n \bmod q$ are separating elements of k/q. If the characteristic is $\neq 0$, it is greater than the degree $[K/p:k/q]$, and so the residue classes of $x_2,\cdots,x_n \bmod p$ are also separating elements of K/p. Let e be the order of ramification of p and z a prime element of p, and $q = z^e u$ with a p-adic unit u. Then

$$\begin{aligned} dx_1\wedge\cdots\wedge dx_n &= \left(\frac{\partial q}{\partial x_1}\right)^{-1} dq\wedge dx_2\wedge\cdots\wedge dx_n \\ &= \frac{ez^{e-1}u + z^e\frac{\partial u}{\partial z}}{\frac{\partial q}{\partial x_1}}\, dz\wedge dx_2\wedge\cdots\wedge dx_n. \end{aligned}$$

We made the assumption at the beginning of §11 that $[K:k]$ is smaller than the characteristic if this is > 0. Then also $e \leq [K:k]$ is smaller than the characteristic and hence $e \neq 0$. So the divisor $(dx_1\wedge\cdots\wedge dx_n)$ is exactly $(e-1)$-times divisible by p. Comparing this with (2) we get the theorem.

At last let p divide y_0. Then we make the projective transformation

$$x_1 = \frac{1}{x_1'}, \quad x_2 = \frac{x_2'}{x_1'}, \quad \cdots \quad , x_n = \frac{x_n'}{x_1'}$$

which exchanges y_0 and y_1. We have

$$dx_1 \wedge \cdots \wedge dx_n = - x_1'^{-n-1} dx_1' \wedge \cdots \wedge dx_n' ,$$

and the contribution of p to $(dx_1 \wedge \cdots \wedge dx_n)$ is $y_0^{-n-1} p^{e-1}$. This completes the proof.

Theorem 2 is a special case of the formula

$$(3) \qquad M(m)_* = y_0^{n+1} M(m)^* = M\left(\frac{1}{m(dx_1 \wedge \cdots \wedge dx_n)}\right) .$$

For the proof we have to refer to the proof of Theorem 1 in §5 where we showed for a reflexive h-module:

$$M(m)^* = \bigcap_p M(m)_p^* .$$

From §10, (4) we deduce

$$M(m)_p = M(1)_p \, m$$

with an $m \in K$ such that $|m|_p = |p_p|^{\nu_p(m)}$ for all p/p (for the existence of such an m see for example: van der Waerden, Algebra II, first chapter). Hence

$$M(m)_p^* = M(1)_p^* \, m^{-1}$$

and $M(1)_p^*$ is the p-adic extension of $M(1)^* = H^*$. Now (3) follows from Theorem 2.

With Theorem 2 and formula (3) we can set up the chief contributions to the Riemann-Roch theorem for the ideals of multiples of a divisor. The rank

$$(4) \qquad L(0, M(m^{-1})) = \dim m$$

is called the *dimension* of M. It has the following meaning: let $m_1, \cdots, m_\ell$ be a basis of the k_0-space of elements of degree 0 in $M(m^{-1})$, namely $\ell = L(0, M(m^{-1}))$. The divisors of all these are

$$(m_i) = \frac{m_i}{m}$$

with integral divisors m_i. The assumption is that all elements $m \in K$ whose divisor is $\frac{n}{m}$ with an integral n are linear combinations of those m_i. According to (3)

$$(5) \qquad L(0, M(m^{-1})_*) = \dim \frac{(dx_1 \wedge \cdots \wedge dx_n)}{m} = \dim m_*.$$

Here

$$(6) \qquad m_* = \frac{(dx_1 \wedge \cdots \wedge dx_n)}{m}$$

is the divisor *canonically conjugate to* m. The definition is bad in the respect that it is based on a special model of K, namely on the variables x_i. But Theorem 1 shows that with a change of the model, the class of $(dx_1 \wedge \cdots \wedge dx_n)$ remains the same. And one can easily see that $\dim m$ and $\dim m_*$ only depend on the divisor class (which is accepted as known or left as an exercise to the reader).

Now the theorem of Riemann-Roch is

$$(7) \qquad \dim m + (-1)^n \dim m_* = \gamma(M(m^{-1})) + \cdots$$

where the dots indicate an expression containing the $\mathrm{Ext}_h^i(M(m^{-1}), h)$. This expression is in general not easily accessible. But if $M(m)$ is a free h-module, it is 0. And we can show in many cases that $M(m)$ is indeed a free module. We note lastly

Theorem 3. *If* $H = M(1)$ *is free,* $\gamma(H)$ *is equal to* $1 + (-1)^n$ *times the number of linearly independent (with respect to* k_0*) differentials* $a\,dx^1 \wedge \cdots \wedge dx^n$ *whose divisors are integral.*

These differentials are called *differentials of 1st kind.*

Proof. For m the unit divisor we have dim (1) = 1. Indeed, here cannot exist an element $m \in K$ the divisor of which is integral. ecause such an m would satisfy an equation

$$m^N + c_1(x)m^{N-1} + \cdots + c_N(x) = 0$$

ith rational functions $c_i(x) \subset k = k_0(x_1, \cdots, x_n)$ whose divisors are lso integral. Now it is clear that the $c_i(x)$ must be constants, and o also m is a constant, and its divisor is the unit divisor.

All divisors in the canonical class are given by the divisor of a ifferential times the divisor of a function, in other words, they are f the form $(adx^1 \wedge \cdots \wedge dx^n)$. This proves the theorem.

In the case $n = 1$ one can calculate $\gamma(M(m^{-1}))$ once $\gamma(H)$ is given. ne only has to know of how many points m consists. For $n > 1$, $(M(m^{-1}))$ is not as easy to describe, and the Riemann-Roch theorem as uch has less explicit applications. But it has played a role in §9. nd, on the other hand, we shall find in §14 that $\gamma(M(m^{-1}))$ has a eaning in connection with the question of the number of common points f n divisors. This teaches us what to expect from the theory when > 1.

12. *Automorphic forms and projective varieties*

We give in §12 a report on the chief facts of the theory of auto- orphic forms which are relevant in our connection. The reader will ind full details in the following paper:

W.L. Baily and *A. Borel*, Compactification of Arithmetic quotients f bounded symmetric domains, Annals of Math. **84**(1966), 442-528.

A shorter presentation of the theory, though under restricted as- umptions is given by

M. Eichler, Zur Begründung der Theorie der automorphen Funktionen in mehreren Variablen, Aequationes Mathematicae 3(1969), 93-111.

Given a complex space **H** of dimension n, more specifically an open subspace of $\mathbf{C}^n$, and a group Γ operating on **H** which has a fundamental domain F. There is a mapping

$$\mathbf{H} \longrightarrow \mathbf{H}/\Gamma = \mathbf{F},$$

a sort of "folding **H** together," which is almost everywhere analytic and even analytically invertible except for the "creases," caused by the fixed points of elements of Γ. The quotient **F** can be identified with a fundamental domain F if boundary points of F are "glued together" in a suitable way.

An *automorphic form* of weight h, $f(z)$, is a holomorphic function in **H** which satisfies the functional equations

$$(1) \qquad f(M(z))\gamma_M(z)^h = f(z) \quad \text{for all} \quad M \in \Gamma.$$

The factor $\gamma_M(z)$ is a holomorphic and never vanishing function in **H** depending on M, and satisfying

$$(2) \qquad \gamma_{MN}(z) = \gamma_M(N(z))\gamma_N(z).$$

We are here particularly interested in a group Γ which is a subgroup of either $\mathrm{Sp}(m,\mathbf{R})$ or of the product $\otimes\, \mathrm{SL}(2,\mathbf{R})$ (n copies). In these cases we consider the substitutions

$$M\colon\ Z \longrightarrow (AZ+B)(CZ+D)^{-1}$$

$$M\colon\ z^\nu \longrightarrow (\alpha^\nu z^\nu+\beta^\nu)(\gamma^\nu z^\nu+\delta^\nu)^{-1} \qquad (\nu = 1,\cdots,n)$$

where $Z = X + iY$ is a complex symmetric matrix with a positive definite imaginary part: $Y \gg 0$ and **H** the space of the $n = \frac{m(m+1)}{2}$ coefficients of Z, or $z^\nu = x^\nu+iy^\nu$ is a set of variables lying in the upper half planes. In these cases possible "automorphic factors" are

$$(3) \qquad \gamma_M(Z) = \det\,(CZ+D)^{-1} \quad \text{or} \quad \gamma(z^\nu) = \prod_{\nu=1}^{n} (\gamma^\nu z^\nu + \delta^\nu)^{-1}.$$

There exist always $n+1$ algebraically independent automorphic forms with respect to such a group, and the ring of J of them is finitely generated. As we have seen in §10, there exist $n+1$ algebraically independent automorphic forms $y_\nu = y_\nu(z)$ of a common weight h, say, such that all automorphic forms depend integrally on

$$h = \mathbf{C}[y_0, \cdots, y_n].$$

We take the y_ν as the (homogeneous) coordinates of the n-dimensional projective space $\mathbf{P}^n$. Now there exists a space $\mathbf{G}$, covering $\mathbf{P}$ in N sheets where N is the degree

$$N = [K:k]$$

of the field K of all *automorphic functions* or quotients of automorphic forms of equal weights, and $k = \mathbf{C}(x_1, \cdots, x_n)$ with $x_i = \dfrac{y_i}{y_0}$.

$\mathbf{G}$ has a natural analytic structure, and it is the 1-1 and bi-analytic image of $\mathbf{F}$, except for two sorts of singular points. The first are the singular points of $\mathbf{F}$, which were mentioned before; these are attached to the fixed points of elements of Γ. The other exceptions are due to the following fact: in many cases F and $\mathbf{F}$ are not compact sets, while $\mathbf{P}$ and then also $\mathbf{G}$ are compact. $\mathbf{G}$ is then the closure of the image of $\mathbf{F}$ in $\mathbf{G}$. This closure or *compactification* of $\mathbf{F}$ is an important, though not simple feature of the theory. Originally it had to be carried out before it could be proved that the ring J of all automorphic forms is finitely generated. In the above quoted paper, the author showed that the finiteness can be proved first, and then the compactification becomes easy.

An important fact in the theory of automorphic forms with factors of the property (3) is the following: Let $z^1, \cdots, z^n$ be either the independent coefficients of the symmetric matrix Z or the variables denoted thus in the other case. Then, if $f_1,\cdots,f_n$ are independent automorphic functions, the Jacobian

$$D_f(z) = \frac{\partial(f_1,\cdots,f_n)}{\partial(z^1,\cdots,z^n)} \tag{4}$$

behaves as an automorphic form of weight $h_0 = m+1$ or 2 in these cases, in other words, $D(z)$ satisfies (1). ($D(z)$ is, of course, not holomorphic.)

If no element $M \in \Gamma$ different from the unit element has an $(n-1)$-dimensional variety of fixed points, we can attach to $D_f(z)$ the divisor

$$(D_f(z)) = (df_1 \wedge \cdots \wedge df_n). \tag{5}$$

Indeed, under this condition, there lies a regular point of $\mathbf{F}$ on (the variety of zeros of) every prime divisor p in the sense of the model given by the y_ν. In the neighborhood of such a point we may choose z^1 as a prime element of p and the other z^i as in the proposition of §11.

This fact leads to a certain symmetry property of the rank function $L(\lambda,J)$ which we now propose to determine. A model h being given, J is split up into the direct sum §11, (6):

$$J = \bigoplus_{r=0}^{h-1} J_r.$$

For large λ rank functions become polynomials $H(\lambda,J_r)$.

For a $f_r \in J_r$, the ideal $f_r^{-1}J_r$ can be described as the H-ideal of multiples defined by the divisor (f_r^{-1}), given by f_r^{-1}:

$$f_r^{-1}J_r = M(f_r^{-1}). \tag{6}$$

Because of (5) and Theorem 2 in §11 we have especially

$$y_0^{n+1}H^* = H_* = M(D_f(z)^{-1}). \tag{7}$$

Applying §11, (3) to the divisor $(f_r)^{-1}$, we get

$$M(f_r^{-1})_* \;=\; M\left(\frac{f_r}{D_f(z)}\right)\left|\frac{\partial(f_1,\cdots,f_n)}{\partial(x_1,\cdots,x_n)}\right| \;\sim\; M\left(\frac{f_r}{D_f(z)}\right).$$

Here $\frac{f_r}{D_f(z)}$ is a "fractional" automorphic form of weight $r-h_0$, and therefore we have the equivalence

$$M(f_r^{-1})_* \sim M(f_{h_0-r}^{-1}) \tag{8}$$

with $f_{h_0-r} \in J_{h_0-r}$. It follows now from the duality theorem of §9 for $y_0^{-\lambda}M$ instead of M and the fact that equivalent modules have the same genus coefficient that

$$\gamma(y_0^{-\lambda}M(f_r^{-1})) \;=\; (-1)^n\gamma(y_0^{\lambda}M(f_{h_0-r}^{-1})). \tag{9}$$

For practical applications we ought to translate this into a formula for the rank functions of J_r. The proposition of §9 says that $\gamma(y_0^{-\lambda}M) = H(\lambda,M)$. We use this for M given by (6). J_r only contains elements of weights $r+\lambda h$. We shall write $H(\lambda,r)$ for the number of linearly independent elements of weight $r+h\lambda$ in J_r, for large λ, and we have

$$H(\lambda,r) \;=\; \gamma(y_0^{-\lambda}M(f_r^{-1})).$$

With this notation (9) can eventually be formulated in the

Theorem. *If no element $M \in \Gamma$ different from the unit element has an $(n-1)$-dimensional variety of fixed points, and if the Jacobian (4) is a "fractional" automorphic form of weight h_0, the rank polynomial $H(\lambda,r)$ of elements of weight $r+\lambda h$ in J_r has the symmetry*

$$H(\lambda,r) \;=\; (-1)^n H(-\lambda,h_0-r)$$

provided that J is a quasifree h-module.

Shimizu's rank formula (see §19) indeed exhibits this symmetry.

§13. *Quasiinvertible ideals*

As was expressed at the end of §11, the theorem of Riemann-Roch does not yield the same satisfaction for $n > 1$ as it does in the case $n = 1$. This is not only due to the Ext^i occuring on one side, but also because the genus coefficient $\gamma(M)$, occurring on the other, is not directly accessible. We shall see in §14 that the easy computability of $\gamma(M)$ for $n = 1$ is an expression of a fact which holds for all n, but which assumes a simple form only for $n = 1$.

Our approach to the problem is by *global*, i.e., ideal theoretic methods, as opposed to *local*, i.e., valuation theoretic ones. However, we do not want to miss the geometric point of view because in the last result we are interested in points of intersections.

§13 is a preparation for §14. Throughout H means the domain of all algebraic forms in $\bar{K} = \oplus Ky_0^\nu$ which depend integrally on h. So H is normal. We are going to apply the method of Chapter I which reduces the number of variables by forming quotient modules and ideals. But this leads us to a more general situation:

Let K be a commutative hypercomplex system with unit element 1, and of rank N over k. In $\bar{K} = \oplus Ky_0^\nu$ a ring H is given whose elements depend integrally on h, which is reflexive, and whose rank is N. We consider H-ideals in $\bar{K}$ which are finite h-modules. Such an ideal M is said to be *quasiinvertible* if another ideal M^{-1} exists such that

$$MM^{-1} \approx H.$$

We may express $M^{-1} = \mathrm{Hom}_H(M,H)$.

We want to characterize quasiinvertible ideals in another way and therefore introduce the following concept: A *place of h* is a homomorphism

$$P\colon h \longrightarrow h'$$

where h' is an integral domain in a graded division ring $\bar{k}'$ of some dimension r. Let f be in the kernel of P. Then after a suitable transformation of the variables we may assume f as normed, and h' can be considered as a h_{n-1}-module. This procedure can be repeated until h'

ecomes a finite h_r-module, and $\overline{k}'$ an extension of the division ring $_r = \bigoplus k_0(x_1,\cdots,x_r)y_0^\nu$. We also introduce the corresponding *local ring*

1) $$h_P = \{\tfrac{a}{b}: a,b \in h,\ b \text{ homogeneous},\ b \longrightarrow b' \neq 0\}$$

b is not mapped on 0 by P) and the local extension

2) $$M_P = h_P M,$$

f an H-ideal, especially $H_P = h_P H$.

Special cases are the maps

$$p: h \longrightarrow h/ph = h'$$

or a prime polynomial p, in which case $r = n-1$. Others are the *oints* P when $r = 1$.

Theorem. If the constant field k_0 is infinite, the local extensions $_P$ of quasiinvertible ideals are principal H_P-ideals.

Conversely, if M_P is a principal H_P-ideal for all points P, M is uasiinvertible.

Remark. We shall see in §15 that all reflexive H_P-ideals are rincipal if P is a regular point (yet to be defined).

Proof. Let M be quasiinvertible and M_ν a system of generators of as h-module, and ξ_ν independent indeterminates. Then the norm of $(\xi) = \sum M_\nu \xi_\nu$ in the sense of the rank equation is a polynomial in the $_\nu$ with coefficients in $\overline{k}$. The g.c.d. of the coefficients is called he *norm* of M: $n(M)$. Because k_0 has ∞ many elements the ξ_ν can be ssigned values $\rho_\nu \in k_0$ such that $M_P = \sum M_\nu \rho_\nu$ has norm $n(M_P) = (M)u_P$ with a unit u_P in the local ring h_P.

Now let M be an arbitrary element in M and $X = \frac{M}{M_P}$. With an in- eterminate ζ, $\zeta - X = \frac{\zeta M_P - M}{M_P}$ has always a norm in $h_P[\zeta]$. But this orm is the rank polynomial. Thus X depends integrally on h_P. All elements of $\overline{K}$ which depend integrally on h_p form a ring $\overline{H}_p$ (which is not a finite h_p-module if $\overline{K}$ is not semisimple). Similarly, we find a $M_p' \in M_p^{-1}$ the norm of which is equal to the norm of M_p^{-1} up to a unit factor, and

$$\frac{1}{M_p} M_p \subseteq \overline{H}_p, \qquad \frac{1}{M_p'} M_p^{-1} \subseteq \overline{H}_p .$$

Because M is quasiinvertible, $M_p M_p^{-1} \approx H_p$ and then

$$\frac{1}{M_p M_p'} H_p \subseteq \overline{H}_p,$$

especially $(M_p M_p')^{-1} \in \overline{H}_p$, and then the norm $n(M_p M_p')^{-1} \in h_p$.

On the other hand, from the definition follows

$$M_p M_p^{-1} \subseteq H_p, \qquad M_p' M_p \subseteq H_p$$

and then

$$M_p M_p' H_p \subseteq H_p.$$

Therefore $M_p M_p'$ must be a unit and

$$M_p M_p^{-1} \cdot M_p' M_p \quad = \quad H_p$$

which can only hold if both factors on the left are H_p. This proves the first part of Theorem 1.

For the second part we form the product $R = MM^{-1}$ with $M^{-1} = \mathrm{Hom}_H(M,H)$. R is an H-ideal contained in H. Let $r = R \cap \overline{K}$. If $r \approx h$, we have also $R \approx H$. Indeed let H_ν be a system of generators of H as an h-module and H an arbitrary element of H. Let $H = \sum H_\nu h_\nu$ with $h_\nu \in h$. If $\deg(H)$ is large, the $\deg(h_\nu)$ are also large and vice-versa. But for large degrees the h_ν are in r due to our assumption, and then $H \in R$. The opposite inclusion is trivial.

If r is not $\approx h$, the quotient ring h/r contains elements of arbitrarily large degrees. Then there exists a homomorphism

$$P: h/\mathfrak{r} \longrightarrow h'$$

where h' is of dimension 0 or, in other words, a finite extension of $k_0[y_0]$ (after a suitable transformation of the variables). But this defines also a point of h.

Now we are assuming that $M_p = H_p M_p$ with an element M_p. So the local extension of MM_p^{-1} at P is H_p. Therefore there exists a polynomial $f \in h$ which is not mapped on 0 by P such that $MM_p^{-1}f \subseteq H$. So $HM_p^{-1}f \subseteq M^{-1}$ and

$$M_p M_p^{-1} f = H_p \subseteq h_p MM^{-1} = h_p R.$$

But then $h_p R = H_p$ and $h_p \mathfrak{r} = h_p$, a contradiction.

In §14 we shall need the following lemma in which we speak of relatively prime ideals. Two integral ideals, M and N, will be called *relatively prime* if for all prime polynomials p:

$$(3) \qquad M_p + N_p = H_p.$$

Lemma. Let M, N *be two integral, quasireflexive and relatively prime* H*-ideals, and assume that* M *is quasiinvertible. Then*

$$M \cap N \approx MN.$$

Proof. For a prime $p \in h$ put $M_p = H_p M_p$, and let $N_{p\nu}$ be a system of generators of N_p as an H_p-ideal. Take an $M = M_p V \in M_p$, with $V \in H_p$. Because of (3) an equation

$$\sum N_{p\nu} U_\nu + M_p W = 1, \qquad U_\nu, W \in H_p$$

holds. It entails

$$MW = V(1 - \sum N_{p\nu} U_\nu).$$

If also $M \in N_p$, V must lie in N_p, and therefore $M \in M_p N_p$. This shows

$$(4) \qquad M_p \cap N_p = M_p N_p ,$$

and because M and N are quasireflexive, the intersections of both sides of (4) are (cf. §5, Theorem 1)

$$(5) \qquad M \cap N = (MN)^{**} .$$

Because M is quasiinvertible, $(MN)^{**}M^{-1} = R$ is an ideal with the property $RM \approx (MN)^{**}$. So the p-adic extensions on both sides are equal, and these are

$$R_p M_p = N_p M_p .$$

Hence $R \subsetapprox N^{**}$ which is $\approx N$ because N is quasireflexive. Therefore $(MN)^{**} \subsetapprox MN$ (contained in the sense of quasiequality), and now follows $(MN)^{**} \approx MN$. This completes the proof of the lemma.

§14. *Intersection numbers*

For n H-ideals M_i we introduce the following formal expression

$$(1) \qquad d(M_1,\cdots,M_n) = \gamma(H) - \sum_{\nu} \gamma(M_\nu) + \sum_{\nu_1<\nu_2} \gamma(M_{\nu_1} M_{\nu_2}) - + \cdots$$
$$+ (-1)^n \gamma(M_1 \cdots M_n).$$

We will call it the *intersection number* of the M_ν. This has of course to be justified.

In the case $n = 1$ we have

$$\gamma(M) = - d(M) + \gamma(H).$$

The theorems 1 and 2 below tell us that $d(M)$ is equal to the number of points lying on M (or on the divisor whose ideal of multiples M is), and that $d(M_0 M_1) = d(M_0) + d(M_1)$. These facts, used on the right hand side of equation (2) in §9, give the theorem of Riemann-Roch its significance. But for $n > 1$, the analogue theorems involve n ideals, nd the situation becomes more complex.

We shall derive the properties of the intersection number inductively. But for the induction on n we need a rather strong

Assumption. If $n > 2$, *the product ideal* $M_1 \cdots M_n$ *and all its partial products occurring in* (1) *are quasifree* h*-modules. In the case of Theorem* 1, *when a further ideal* M_0 *is given, the same is assumed for the product* $M_0 M_1 \cdots M_n$ *and its partial products.*

For $n \leq 2$ no such assumption is required, due to Theorem 1 in §9.

In general the assumption raises a number of questions: Can it be expressed as an invariant property of the underlying variety? Is it always satisfied in the case of a regular variety? Does it follow from the simpler assumption that only H is a quasifree h-module? One may expect positive answers, but we do not take up the questions.

The rôle played by our assumption is this: from Theorem 3 of §8 follows that all these ideals are quasireflexive h-modules. We need that for the lemma of §13. And by Theorem 2 of §8 the property of quasifreeness is inherited by the quotients.

Two properties of the intersection number are evident from the definition: $d(M_1, \cdots, M_n)$ is symmetric in all arguments, and it remains unchanged if the M_i are replaced by equivalent ideals $A_i M_i$ with $A_i \in K$ (degree 0) and not divisors of zero. Namely then $\gamma(A_i M_i) = \gamma(M_i)$. The following proposition provides us with the tool for the derivation of the other properties.

Proposition. If M_n *is an integral ideal and as an* h*-module of the same rank as* H, *a transformation of the variables is possible such that* $H' = H/M_n$ *is a ring in a hypercomplex system* $\overline{K'}$ *over* $\overline{k}_{n-1}$ *with unit element. The* $M_i' = M_i/M_i M_n$ *are* H'*-ideals, and*

$$d(M_1,\cdots,M_n) \quad = \quad d(M_1',\cdots,M_{n-1}').$$

Proof. We only have to be sure that $fM_n \subseteq H$ with a normed f, and then H' is a torsion free h_{n-1}-module. The rest is evident because of equation (6) in §6.

Lemma. If M_i have the same rank as H and $M_2,\cdots,M_n$ are reflexive and quasiinvertible, there exist elements $A_i \in \overline{K}$ which are not divisors of zero, such that $B_i = A_iM_i \subseteq H$ and that the quotient

$$H/(B_1+\cdots+B_n)$$

is a finite h_0-module, after a suitable transformation of the variables

Remark. The statement of the lemma includes that any two of the B_i are relatively prime.

Proof. The statement is evident for $n = 1$, and is being taken as true for n-1. We choose at first $A_1,\cdots,A_{n-1}$ only so that they are not divisors of 0, and that $A_iM_i = B_i \subseteq H$. After this we seek A_n such that

$$(2) \qquad B_{\nu p} + B_{np} \quad = \quad H_p$$

for all $p \in h$. This is possible for the following reason. Because of the theorem in §13, $M_{np} = H_pM_p$ with M_p not a divisor of 0. Now almost all $B_{\nu p} = H_p$. There exists an $A_n \in \overline{K}$ such that A_nM_p is a p-adic unit for all exceptions and $A_nM_p \in H_p$ for all other p.

With A_n and B_n so constructed we consider

$$H' \quad = \quad H/B_n, \qquad B_\nu' \quad = \quad B_\nu/(B_\nu \cap B_n).$$

The B_ν' are H'-ideals of maximal rank. We claim that, for $\nu \geq 2$, they are quasiinvertible, Let $B_\nu B_\nu^{-1} \approx H$. There exists an $f \in h$ such that $fB_\nu^{-1} \subseteq H$. When constructing A_n we can arrange that B_n also is elatively prime to fB_ν^{-1} in the sense of (2). Then

$$B_\nu/B_\nu B_n \cdot fB_\nu^{-1}/fB_\nu^{-1}B_n \approx fH/fB = f'H'$$

here f' is the residue of f mod $h \cap B_n$. H' and the B_ν' ($\nu \geq 2$) are uasireflexive due to the assumption and Theorem 3 in §8. Let H'', B_ν'' be heir reflexive completions. Now the induction assumption gives us lements A_ν', not divisors of 0, such that $B_\nu''' = A_\nu' B_\nu'' \subseteq H_\nu''$ and $''/(B_1''' + \cdots + B_{n-1}''')$ is a finite h_0-module, after a suitable transformaion of the variables. Then also $A_\nu' B_\nu'$ and H' have the same properties ecause these are quasireflexive modules. If A_ν are suitable elements f $\bar{K}$ which are mapped on the A_ν' by this formation of residues we have ventually

$$H/(A_1B_1 + \cdots + A_{n-1}B_{n-1} + B_n) \cong H'/(A_1'B_1' + \cdots + A_{n-1}'B_{n-1}'),$$

ith which the lemma is proved.

Theorem 1. *Let* $n+1$ *H-ideals* $M_0, \cdots, M_n$ *be given which, as h-odules, are of maximal rank in* $\bar{K}$ *and quasireflexive if* $n \leq 2$ *or satsfying the assumption at the beginning of* §14 *if* $n > 2$*, and which are lso quasiinvertible with one possible exception. Then*

$$d(M_0M_1, M_2, \cdots, M_n) = d(M_0, M_2, \cdots, M_n) + d(M_1, M_2, \cdots, M_n).$$

Proof. If necessary we exchange M_0 and M_1 such that M_0 is quasinvertible. We take $A_1, \cdots, A_n$ according to the lemma and also A_0 uch that $B_0 = A_0M_0$ is $\subseteq H$ and relatively prime to B_n.

We prove the theorem first for the B_i instead of the M_i.

For $n = 1$ we use §6, Proposition 2. In this case the theorem tates because of §5, Proposition 2

$$G(H) - G(B_0) = G(B_1) - G(B_0B_1),$$

hich is easily verified. Now we assume the theorem true for n-1. We may replace the B_i by their reflexive completions; because they are quasireflexive, the genus coefficients remain the same. For $n = 1$ we need not make more assumptions as that B_0 is quasiinvertible.

We introduce

$$H' = H/B_n, \quad B'_\nu = B_\nu/(B_\nu \cap B_n) \approx B_\nu/(B_\nu B_n)$$

(because B_ν or B_n is quasiinvertible, the lemma of §13 gives us

$$B'_\nu = B_\nu/B_\nu B_n \approx B_\nu/(B_\nu \cap B_n).$$

The B'_ν are quasireflexive h_{n-1}-modules of maximal ranks, and all but one quasiinvertible H'-ideals (see the proof of the preceding lemma). Now the above proposition furnishes the induction.

After the theorem is proved for the $B_i = A_i M_i$ we must extend it to the M_i. Let α_i be the degrees of the A_i such that $a_i = A_i y_n^{-\alpha_i}$ have degrees 0. As was stated at the beginning, multiplication by a_i does not alter the genus coefficients. So our last task is to show: if the theorem is true for some M_i, it is also true if one of the M_i is multiplied or divided by y_n. This is a formal computation.

As easily checked, the formula of the theorem can be expressed as

$$(3) \qquad d_{n+1}(M_0, M_1, \cdots, M_n) = \gamma(H) - \sum_{\nu=0}^{n} \gamma(M_\nu) + - \cdots = 0,$$

where the alternating sum is now taken for $n+1$ arguments. Furthermore

$$(4) \qquad d_{n+1}(y_n M_0, M_1, \cdots, M_n) - d_{n+1}(M_0, M_1, \cdots, M_n)$$
$$= - d_{n+1}(M'_0, M'_1, M'_2, \cdots, M'_n) + d_n(M'_1, M'_2, \cdots, M'_n)$$

with

$$M'_i = M_i/y_n M_i.$$

We have to prove if $d_{n+1}(M_0, \cdots) = 0$, then also $d_{n+1}(y_n M_0, \cdots) = 0$, and vice versa. This is again done by induction on n. For $n = 1$ the theorem was already proved. For the induction we can see that both terms on the right of (4) vanish. This completes the proof.

The next theorem establishes a connection between the global and the local properties of the variety attached to H. For this we need a preparation. We assume the constant field k_0 as algebraically closed. A *point* of H or of the corresponding variety is a homogeneous homomorphism

$$P: H \longrightarrow k_0[y_0].$$

This definition suffices for the present purpose. In §15 we shall discuss the properties of points in more detail.

Theorem 2. Let n integral H-ideals $M_1,\cdots,M_n$ be given; they are all assumed quasiinvertible and quasireflexive h-modules if $n \leq 2$, and even quasifree if $n > 2$.

Assume further that $H/(M_1+\cdots+M_n) = H_0$ is, after a suitable transformation of the variables, a finite h_0-module. Let H_{01} be the torsion-free kernel of H_0. Then $\overline{K}_0 = \overline{k}_0 H_{01}$ is the direct sum of graded primary rings $\overline{K}_{0i}$. Let $\overline{R}_{0i}$ be the radical of $\overline{K}_{0i}$. Then there exists a unique homomorphism

$$\overline{K}_0 \longrightarrow \overline{K}_{0i}/\overline{R}_{0i} \cong \overline{k}_0 = \bigoplus_{\nu=-\infty}^{+\infty} k_0 y_0^{\nu}.$$

for every i, and its restriction to H_0 yields a homomorphism

$$P_i: H \longrightarrow H_0 \longrightarrow k_0[y_0].$$

To each of these P_i attach the intersection multiplicity

$$\delta(M_1,\cdots,M_n;P_i) = [\overline{K}_{0i}:\overline{k}_0].$$

Then the total number of intersections is

$$\sum_i \delta(M_1,\cdots,M_n;P_i) = d(M_1,\cdots,M_n).$$

Remark 1. The multiplicities can also be expressed as

$$[K_{0i}:k_0] = [\overline{K}_{0i}:\overline{k}_0]$$

where K_{0i} is the set of elements of degree 0 in $\overline{K}_{0i}$. It is easy to show that they are invariants of the underlying abstract variety and of the divisors m_i whose ideals of multiples are $M(m_i) \approx M_i$. But first the meaning of the abstract variety attached to H would have to be defined. We shall not do this here.

In the case $n = 2$, *O. Zariski* (An introduction to the theory of algebraic surfaces, Springer Lecture Notes No. 83; p. 67) gives another proof of Theorem 2. He assumes the variety to be non-singular. In that case all quasireflexive H-ideals are quasiinvertible, as we shall show in §15.

Remark 2. The simplest case of Theorem 2 is the theorem of Bézout. Let $H = h$ and $M_\nu = hM_\nu$ with polynomials M_ν of degrees μ_ν. An easy calculation gives

$$d(M_1,\cdots,M_n) = \mu_1\cdots\mu_n.$$

So this is the number of common zeros of the M_ν.

Theorem 2 can easily be checked in this case.

Proof by induction on n. For $n = 1$ we have seen in the proof of Theorem 1 that (here we need that M_1 is quasiinvertible)

$$d(M_1) = G(M_1) - G(H).$$

The rank of the torsion-free kernel of H/M_1 over h_0 is $G(M_1) - G(H)$ according to §5. And this in turn is equal to the sum of the ranks $[\overline{K}_{0i}:\overline{k}_0]$.

For $n > 1$ we form again ($\nu < n$)

$$H' = H/M_n, \quad M'_\nu = M_\nu/(M_\nu \cap M_n) \approx M_\nu/M_\nu M_n.$$

he last $\approx$ follows from the lemma in §13, because M_ν is quasiinvertible, nd any two of the M_ν are relatively prime because otherwise $/(M_1+\cdots+M_n)$ could not be a finite h_0-module. Now we have

$$H/(M_1+\cdots+M_n) \cong H'/(M_1'+\cdots+M_{n-1}'),$$

nd the theorem follows from the induction assumption by the proposi- ion.

15. *Regular local rings*

In §15 we assume $n = 2$. We will prove a theorem which we shall eed in the last chapter. We could very well take this result from iterature, and even more so since we shall use quite a few facts in hapter III for which we must refer to other sources. But the general heory developed so far gives us the tools for a new proof of this heorem, and even for a certain extension of it. Most of the following onsiderations are valid for an arbitrary n.

In §15 we also assume J to satisfy the conditions of §10 with an rbitrary constant field k_0, and J is also assumed normal.

In analogy to §13 we define a *place* of J a homomorphic map

$$P: J \longrightarrow \overline{J}$$

n a graded integral domain $\overline{J}$ of a dimension $r < n$. The number r is he *dimension* of the place. To P we attach the *graded local ring*

$$J_P = \{\tfrac{a}{b}:\ a, b \text{ (homogeneous)} \in J,\ b \xrightarrow[P]{} \overline{b} \neq 0\}. \tag{1}$$

t has a maximal ideal P_P consisting of all such quotients whose numer- tors are mapped on 0.

To J_P there corresponds the non-graded *local ring*

$$J_P^0 = J_P \cap K.$$

rom J_P^0 one can easily retrieve

$$J_P = J_P^0 J,$$

and properties of J_P^0 can be translated into properties of J_P, and vice versa. In the rule, non-graded local rings are used, but we restrict ourselves to graded ones.

It may be noted that there exist more general local rings than those attached to our graded local rings. These cannot be treated by our methods.

A graded local ring J_P is called *regular* if there exist

$$s = n - r \tag{2}$$

elements $p_1, \cdots, p_s$ of equal degrees such that

$$P_P = J_P p_1 + \cdots + J_P p_s \tag{3}$$

and that there exists also a unit in J_P of the same degree as the p_i.

Examples of places are the valuations of rank 1 or the prime divisors $P = p$. In this case

$$J_p = \{a \in \overline{K}: |a|_p \leq 1\}, \quad P_p = \{a \in \overline{K}: |a|_p < 1\},$$

and J_p is always regular. Places of degree 0 are called *points*.

A torsion-free J_P-module M_P is called *reflexive* if

$$M_P^{**} = M_P, \quad \text{where} \quad M_P^* = \mathrm{Hom}_{J_P}(M_P, J_P).$$

Theorem 1 of §5 can immediately be extended to J-modules; relatively prime elements are those whose divisors are relatively prime. The concept of relatively prime elements is carried over to J_P in a natural way, and so this theorem remains true in the local case.

We shall prove the following

Theorem 1. *If* $n = 2$, *a torsion-free and reflexive module for a regular graded local ring* J_P *which as no divisors of* 0 *is a free module.*

Corollary. Such a local ring is a unique factorization domain.

The corollary is an immediate consequence of the theorem. It is ue to Auslander and Buchsbaum and usually proved in a different way see for instance O. Zariski and P. Samuel, Commutative Algebra, Vol. I, Appendix 7), although there is some relationship between both roofs. We shall reach our goal through a number of lemmas, but we bein with constructing a special model particularly fit for our purpose.

Let $p_1,\cdots,p_s$ be the elements in (3) and $q_0,\cdots,q_r$: $r+1$ eleents in J_P of the same degrees as the p_i the residues mod P_P of hich are algebraically independent. Then there exists a polynomial $\in h$, not mapped on 0 by P, for which

$$y_0 = q_0 f, \quad \cdots, \quad y_r = q_r f, \qquad y_{r+1} = p_1 f, \quad \cdots, \quad y_n = p_s f$$

re lying in J. Let J' be the domain of elements of J which depend ntegrally on

$$h \quad = \quad k_0[y_0,\cdots,y_n].$$

imilarly to (1) we form the local ring J'_P and claim that

$$J'_P \quad = \quad J_P. \tag{4}$$

For the proof we start from the characterization of J_P as the doain of all $w \in \bar{K}$ satisfying an equation

$$F(w;q_0,\cdots,q_r;p_1,\cdots,p_s) \quad = \quad c_0(q_0,\cdots;p_1,\cdots)w^N + \cdots$$

$$+ \; c_N(q_0,\cdots;p_1,\cdots) \quad = \quad 0$$

here the q_ρ are homogeneous elements of J_p which are algebraically inependent mod P_P, and the $c_i(q_0,\cdots;p_1,\cdots)$ homogeneous polynomials, nd where at last $c_0(q_0,\cdots;0,\cdots,0) \neq 0$. This is so because J has een assumed as normal, and therefore also J_P is integrally closed in . It amounts to the same if the q_i and p_i are replaced by the $q_i f$ and the $p_i f$, and then (4) is clear.

We shall also use the

$$h_\nu = k_0[y_0,\cdots,y_\nu] \quad \text{for} \quad \nu = n, n-1,\cdots,r+1$$

and their "localizations"

$$h_{\nu,P} = \{\tfrac{a}{b}:\ a,b \in h_\nu,\ b \xrightarrow[P]{} \bar{b} \neq 0\}.$$

(For $\nu = n$ the subscript is being omitted as usual.) These rings have already been discussed in §7, where they were denoted by h_ν^r. We recollect that $h_{\nu-1,P}$ represents the residues of $h_{\nu,P} \bmod y_\nu h_{\nu,P}$ in a unique way.

Lemma 1. *If* M *is a finite, torsion-free, and reflexive* h*-module,* $M_P = h_P M$ *is a* h_P*-module with these same properties.*

If M_P *is a finite, torsion-free, and reflexive* h_P*-module, there exists a* h*-module with these properties and such that* $M_P = h_P M$.

Proof. The first statement is evident. For the second let M_ν be a system of generators of M_P. The M_ν generate a finite and torsion-free h-module N, of which we form the reflexive completion.

$$M = \bigcap_p N_p = N^{**}.$$

Now for all prime polynomials $p \in h$ which vanish in P:

$$h_{P,p} M_p = h_{P,p} N_p = M_{P,p}.$$

So the p-adic extensions of M_P and $h_P M$ coincide, and because M_P and $h_P M$ are reflexive, these modules are identical.

Lemma 2. *For a finite, torsion-free, and reflexive* h*-module* M *we have*

$$\mathrm{Hom}_{h_P}(h_P M, h_P) = h_P \,\mathrm{Hom}_p(M,h)$$

and

$$\mathrm{Ext}^i_{h_P}(h_P M, h_P) = 0$$

for $i > 1$. (Here we use $n = 2$.)

Proof. We write the first equation in easily understandable abbreviation, thus

$$M_P^* = h_P M^*.$$

The inclusion

$$M_P^* \supseteq h_P M^*$$

is trivial. Let $M_P^* \in M_P^*$. Then $M_P^* M \subseteq h_P$. Taking into account the formation of h_P by quotients we find now a $f \in h$ which is a unit in h_P, and for which $f M_P^* M \subseteq h$ holds. This means $f M_P^* \in M^*$, and because f is a unit in $h_P, M_P^* \in h_P M^*$. This gives us the first assertion.

For the second take a free resolution $\cdots \longrightarrow M^0 \longrightarrow M \longrightarrow 0$ of M and the corresponding sequence

$$\cdots \longleftarrow M^{1*} \underset{\mu_1^*}{\longleftarrow} M^{0*} \longleftarrow 0.$$

Write

$$M_{00}^{i*} = M^{i-1,*}\mu_i^*, \quad M_0^{i*} = \ker \mu_{i+1}^* \text{ in } M^{i*}.$$

Then $\mathrm{Ext}^i_h(M,h) = M_0^{i*}/M_{00}^{i*}$. Now the $h_P M^i$ form a free resolution of M_P, and because of the first part of the lemma, $M_P^{i*} = h_P M^{i*}$. This gives us

(5) $$M_{P00}^{i*} = M_P^{i-1,*}\,\mu_i^* = h_P M_{00}^{i*}$$

and

(6) $$M_{P,0}^i = \ker \mu_{i+1}^* \text{ in } M_P^{i*} = h_P(\ker \mu_{i+1}^* \text{ in } M^{i*}).$$

Because of Theorem 1 in §9 $\mathrm{Ext}^i_h(M,h)$ is annihilated by all polynomials of sufficiently large degrees. This means

$$fM_0^{i*} \subseteq M_{00}^{i*}.$$

Taking for f a unit of h_P we find from (5) and (6):

$$M_{P,0}^{i*} = M_{P,00}^{i}$$

or $\mathrm{Ext}_{h_p}^{i}(M_P, h_P) = 0$.

Lemma 3. *Let $\overline{k}_\nu$ be the ring of quotients of h_ν and $h_{\nu,P}$. Let the element $a \in \overline{k}_\nu$ be expandable into a y_ν-adically convergent series*

$$a = \sum_{i=0}^{\infty} \alpha_i y_\nu^i \quad \textit{with} \quad \alpha_i \in h_{\nu-1,P}.$$

Then, if $\nu > r$, $a \in h_{\nu,P}$.

Proof. Let

$$a = \frac{u}{v}, \quad v = v_0 - wy_\nu,$$

and

$$u = u_0 + u_1 y_\nu + \cdots + u_m y_\nu^m, \quad w = v_1 + v_2 y_\nu + \cdots + v_m y_\nu^m$$

with coefficients $u_i, v_i \in h_{\nu-1,P}$. $h_{\nu,P}$ is a Noetherian ring, and v can be decomposed in finitely many prime factors (not necessarily in a unique way). We prove the lemma by induction on the number of prime factors of v, and it is obvious that only the beginning of the induction must be taken care of.

So assume v a prime element of $h_{\nu,P}$. If the v_i have a common divisor t in $h_{\nu-1,P}$ we have $v = t\left(\frac{v_0}{t} - \frac{w}{t} y_\nu\right)$ where the second factor is a unit. Then also $\frac{v_0}{t}$ is a unit, and $\left(\frac{v_0}{t} - \frac{w}{t} y_\nu\right)^{-1}$ allows a y_ν-adic expansion with coefficients in $h_{\nu-1,P}$. So it only needs to treat the case $v = t$, but then the assertion is obvious.

So we now assume that the v_i have no common divisor $t \in h_{\nu-1,P}$. If v_0 is a unit, also v is a unit, and nothing to prove. But if v_0 is not a unit, we have

$$a = \frac{u}{v} = \frac{u}{v_0}\left(1 + \frac{w}{v_0} y_\nu + \left(\frac{w}{v_0}\right)^2 y_\nu^2 + \cdots\right).$$

Here we insert the polynomial expressions for u and w and obtain a y_ν-adic expansion with coefficients in $k_{\nu-1}$ whose coefficients have infinitely increasing denominators unless only finitely many powers of y_ν occur. But since such an expansion is uniquely determined, and since we assume the coefficients in $h_{\nu-1,P}$, the series must terminate after finitely many steps. QED

Lemma 4. *A finite, torsion-free, and reflexive h_P-module M_P is free.*

Proof by induction on n, similar to the proof of Theorem 4 in §8. The beginning is at $n = r+1$ (or $s = 1$). Then all torsion-free graded modules are free. For the induction we use the reduction lemma of §7 in the long exact sequence attached to

$$0 \longrightarrow y_n M_P^* \longrightarrow M_P^* \longrightarrow M_P'' = M_P^*/y_n M_P^* \longrightarrow 0$$

and find from Lemma 2 that

$$M_P' = M_P/y_n M_P \cong y_n^{-1}(M_P'')^*,$$

so M_P' is again finite, torsion-free, and reflexive, and hence free as a $h_{n-1,P}$-module.

Let M_i' be a basis of M_P' and M_i elements of M_P whose residue classes mod $y_n M_P$ are the M_i'. An element $M \in M_P$ can now be expressed as

$$M = \sum_i \alpha_i M_i + y_n M^{(1)} \quad \text{with} \quad \alpha_i \in h_{\nu-1,P},\ M^{(1)} \in M_P.$$

$M^{(1)}$ is expressed in the same way, and so forth. This leads to y_n-adically convergent series a_i of the sort considered in Lemma 3 for which

$$M = \sum_i a_i M_i.$$

So far the coefficients a_i lie in the perfect completion of $\bar{k}$. But because M and the M_i lie in M, and the M_i are linearly independent, the a_i lie even in $\bar{k}$. Because of lemma 3 these a_i are lying in h_P, and the lemma is proved.

In the proof of our theorem we have to show the analogue for J_P instead of h_P. We will also carry out induction on n and for this reason introduce the residue rings

$$J_{n-1,P} = J_P/y_n J_P, \ldots, J_{r+1,P} = J_{r+2,P}/y_{r+2}J_{r+2,P}.$$

Lemma 5. *These $J_{\nu,P}$ are finite, torsion-free, and reflexive $h_{\nu,P}$-modules. $J_{r+1,P}$ is a "graded principal ideal domain" with the only prime ideal $y_{r+1}J_{r+1,P}$.* (Graded principal ideal domain means that every graded ideal is principal.)

Proof. This is almost clear. For the torsion-freeness see §5, Theorem 2. For the reflexivity see the proof of Lemma 4. That $J_{r+1,P}$ is a "graded principal ideal domain" follows from the same fact for $h_{r+1,P}$, because both rings have the same prime element y_n.

Lemma 6. *A finite and torsion-free J_P-module M_P is a reflexive J_P-module, if and only if it is a reflexive h_P-module.*

Proof. This follows by using the second criterion for reflexivity in §5, Theorem 1 for M_P as a J_P-module and as a h_P-module.

Proof of the theorem. Because of Lemma 6 we may replace the assumption that M_P is J_P-reflexive by the other that it is h_P-reflexive. This is easier to handle. As usual we proceed by induction on n, beginning with $n = r+1$. Then, as mentioned in Lemma 5, J_P is a graded principal ideal domain, and any graded module is free.

In the induction argument we shall use a basis b_i of J_P as h_P-module; such a basis exists because of Lemma 4.

We form $M'_P = M_P/y_n M_P$. As in the proof of Lemma 4 we see that it is a finite, torsion-free, and reflexive h_P-module. Of course it is also a $J_{n-1,P}$-module. By the induction assumption it is free. Let M'_i be a basis, and $M_i \in M_P$ whose residues mod $y_n M_P$ are the M'_i. M_P and M'_P have the same dimensions, and the M_i are linearly independent over h_P. The elements $M_i b_j$ form a basis of the space $\bar{k}M_P$ over $\bar{k}$. Their residues mod $y_n M_P$ form a basis of M'_P with respect to $h_{n-1,P}$. Therefore, by the argument of the proof of Lemma 4, the $M_i b_j$ form an h_P-basis of M_P. So the M_i form a J_P-basis of M_P. With this the proof is ready.

A consequence of the corollary of Theorem 1 and the theorem in §13 is

Theorem 2. If a divisor $\mathfrak{p}$ *does not contain a singular point* P*, in other words, if the homonorphism* $\mathfrak{p}: H \longrightarrow H_{\mathfrak{p}}$ *cannot be continued to such a* $P: H \longrightarrow H_P$*, the ideal* $M(\mathfrak{p})$ *of multiples of* $\mathfrak{p}$ *is quasiinvertible.*

CHAPTER III

APPLICATIONS TO MODULAR FORMS

§16. *Introduction*

From now on we restrict ourselves to the Hilbert and Siegel modular forms, and we shall briefly distinguish the cases [H] and [S].

[H] Ω is a totally real algebraic number field of degree $[\Omega:\mathbf{Q}] = n$. The n conjugates of an $\alpha \in \Omega$ will be denoted by α^i. Let o be the principal order. Then

$$\Gamma = SL(2,o) = \{\begin{pmatrix} \alpha & \beta \\ \gamma & \delta \end{pmatrix} : \alpha,\beta,\gamma,\delta \in o \quad \text{and} \quad \alpha\delta - \beta\gamma = 1\}$$

is the Hilbert modular group. Modular forms of weight h are holomorphic functions in a set $z = (z^1,\cdots,z^n)$ of variables restricted to the upper half-plane, and which satisfy the functional equations

$$f(M(z))\left(\prod_{i=1}^{n} \frac{1}{\gamma^i z^i + \delta^i}\right)^h = f(z), \qquad M \in \Gamma$$

where

$$M(z) = (M^1(z^1),\cdots,M^n(z^n)), \qquad M^i(z^i) = \frac{\alpha^i z^i + \beta^i}{\gamma^i z^i + \delta^i}.$$

The domain of the variables z^i is called $\mathbf{H}$.

[S] $Z = X + iY$ is a symmetric m-rowed matrix with positive definite imaginary part. The modular group is

$$\Gamma = Sp(m,\mathbf{Z}) = \{M: M^t \begin{pmatrix} 0 & I \\ -I & 0 \end{pmatrix} M = \begin{pmatrix} 0 & I \\ -I & 0 \end{pmatrix}, \text{ rat. int. coeff.}\}$$

The number of rows of M is $2m$, and M is always subdivided into 4 m-rowed matrices:

$$M = \begin{pmatrix} A & B \\ C & D \end{pmatrix}.$$

he action on the variable matrix is

$$M(Z) = (AZ+B)(CZ+D)^{-1},$$

nd a modular form of weight h is a holomorphic function in the $n =$ $m(m+1)$ variable elements of Z satisfying the functional equations

$$f(M(Z))\,|CZ+D|^{-h} = f(Z), \quad M \in \Gamma.$$

he domain of the variable matrix is called **H**.

In the special case $n = 1$ or $m = 1$ we get the classical modular orms in one variable (we shall always style them as "classical" in rder to avoid the rather ambiguous word "dimension"). But in this case ur definition is not yet complete. Indeed, assumptions on the behav- our in the so-called cusps of the fundamental domain are necessary. t is a deep, though not complicated fact that such assumptions are uperfluous for $n > 1$. But we are only interested in the theory for > 1, while we consider the theory for $n = 1$ as known (although we ctually only make use of a few isolated facts of it).

In §12 we saw that there exists a system of n+1 algebraically inde- endent variables $y_0,\cdots,y_n$ of equal weights such that the ring J of ll modular forms depends integrally on

$$h = \mathbf{C}[y_0,\cdots,y_n].$$

uch a system will be called in the sequel an *admissible coordinate ystem*. The common weight of the y_ν will be denoted by h.

Now we propose the following

Hypothesis. There exists an admissible coordinate system h such hat the ring J of all modular forms is a free h-module.

The same will then also be the case for the submodules of J whose weights lie in a given residue class mod h. Particularly we shall have to use the subring H whose elements have weights divisible by h.

The hypothesis is true in a number of individual cases when J has been explicitly constructed. The case [S] with $m = 2$ was first treated by *Igusa*. The latest and easiest proof was given by *E. Freitag*: Zur Theorie der Modulformen zweiten Grades, Nachr. Akad. Wiss. Göttingen, II, math.-phys. Kl, 1965. For similar cases see *E. Freitag*: Modulformen zweiten Grades zum rationalen und Gausschen Zahlkörper, Sitz.-Ber. Heidelberger Akad. Wiss., math.-phys. Kl, 1967. Similarly *K.B. Gundlach* considered some cases [H] with $n = 2$: Die Bestimmung der Funktionen zu einigen Hilbertschen Modulgruppen, Journ. reine angew. Math. **220**(1965), 109-153.

A general argument in favour of our hypothesis is the following. *J.P. Serre* showed that the ranks $L(0,\eta^{n+1}\mathrm{Ext}_h^i(H,h))$ are equal to the ranks of the homology groups $H^{n-i}(X,O)$ of the underlying variety X in the structure sheaf O (Faisceaux algébriques cohérents, Annals of Math. **61**(1955), 197-278). In the case of varieties without singular points *P. Dolbeault* proved further that the latter ranks are equal to the ranks of the $(n-i)$-th homology groups of the holomorphic differentials

$$\omega^{n-i} = \sum u_\nu \, dx_{\nu_1} \wedge \cdots \wedge dx_{\nu_{n-i}}$$

(Sur la cohomologie des variétés analytiques complexes, C.R. Acad. Paris **236**(1953), 175-177). Our varieties X have always singular points. Nevertheless one may conjecture that the latter equality still holds. In the case of automorphic forms, defined in the product of upper half-planes, *Y. Matsushima* and *G. Shimura* showed that no holomorphic differential forms of degrees $0 < i < n$ exist (On the cohomology groups attached to certain vector valued automorphic forms on the product of upper half-planes, Annals of Math. **78**(1963), 418-449). Their proof assumes the fundamental domain of the group to be compact. But it can vidently be generalized, e.g., to Hilbert modular forms. Assuming olbeault's rank equation we would now get $L(0,\eta^{n+1}\mathrm{Ext}^{i}_{h}(H,h)) = 0$ for > 1. If J is also known as quasifree (§8), we conjecture that from he vanishing of these ranks for one admissible coordinate system h here follows the existence of another one h' such that all these $\mathrm{xt}^{i}_{h'}(H,h') = 0$, and then H is free (Theorem 4 in §8).

We have vainly tried to prove our hypothesis. Naturally the arithetical nature of the modular groups must somehow be linked with the lgebraic properties of the ring of modular forms.

In the following we derive some consequences of the hypothesis. he first is analogue of the theorem of *Appell* and *Humbert* in the theory f theta functions (§18). In §19 we consider some arithmetically deined curves in the "Hilbert modular plane" and compute their intersecion numbers in two ways. Also in this connection our hypothesis plays rôle. The chief algebraic tools are the reduction lemma of §7 and he theorems of §14 on the intersection number.

17. *Specializations of modular forms*

Such specializations have been studied by many authors. We will o the same, and in §17 put together the necessary function theoretical aterial.

By putting all variables $z^{\nu} = \zeta$, equal to one variable ζ, we obain a homomorphic mapping of the ring J of Hilbert modular forms into he ring of classical modular forms in ζ. The weights of the forms are ultiplyed by n. Let J_1 be the image of J under this specialization. e cannot decide in general which subring of all classical modular orms this image J_1 is. But if we know that J_1 is already the full ing, we can trace back properties of J_1 to J.

Let us consider particularly the case $n = 2$. A special Hilbert odular form which can always be constructed is the *Eisenstein series*

$$G_k(z^1,z^2) = \sum \frac{1}{((\mu^1 z^1+\nu^1)(\mu^2 z^2+\nu^2))^k},$$

extended over all pairs $\mu,\nu \in \mathfrak{o}$, different from 0, 0, but in each class $(\varepsilon\mu,\varepsilon\nu)$ of associated pairs (with ε units of $\mathfrak{o}$ of norm 1) only one pair has to be taken. Inserting $z^1 = z^2 = \zeta$ one gets a classical modular form of weight 4. This does not vanish, namely for $\zeta \longrightarrow i\infty$ we get

$$G_2(i\infty,i\infty) = \sum \frac{1}{n_{\Omega/Q}(\nu)^2}$$

which is > 0. It is known that there exists but one modular form of weight 4 up to a constant factor; it is called $g_2(\zeta)$. So $G_2(\zeta,\zeta) = cg_2(\zeta)$.

Let us assume that the similar Eisenstein series

$$G_3(z^1,z^2) = \sum \frac{1}{(\mu^1 z^1+\nu^1)^3(\mu^2 z^2+\nu^2)^3}$$

is mapped on a non-vanishing classical modular form of weight 6. This is the case if

$$G_3(i\infty,i\infty) = \sum \frac{1}{n_{\Omega/Q}(\nu)^3} \neq 0. \tag{1}$$

This is only possible if the basic unit ε of Ω has norm +1, otherwise $n(\nu)$ and $n(\varepsilon\nu)$ cancel out against each other. Our inequality can more or less easily be checked in individual cases. For instance one knows it to hold for $\Omega = \mathbf{Q}(\sqrt{3})$. Under this condition the classical theory tells that $G_3(\zeta,\zeta) = cg_3(\zeta)$, and $g_2(\zeta)$ and $g_3(\zeta)$ generate the whole ring of classical modular forms.

A similar specialization, namely $Z = I\zeta$ with I the unit matrix is possible for Siegel modular forms. Then it is even more difficult to find out which subring of classical modular forms is obtained by this specialization. But there is another specialization which has first been used by Siegel in his introduction into the theory (1939).

Partition the m-rowed variable matrix

$$Z = \begin{pmatrix} Z_1 & z \\ z^t & \zeta \end{pmatrix}$$

here Z_1 is $(m-1)$-rowed, z a $(m-1)$-column vector, z^t its transpose and a scalar variable. Because a Siegel modular form $f(Z)$ is invariant nder $\zeta \longrightarrow \zeta+1$ it allows a Fourier expansion

$$f(Z) = \sum_{n=0}^{\infty} \phi_n(Z_1,z)\, e^{2\pi i n\zeta}.$$

o terms with $n < 0$ occur which we must take without proof. We apply he transformations

$$Z \longrightarrow M(Z) = (AZ+B)(CZ+D)^{-1}$$

or symplectic matrices

$$M = \begin{pmatrix} A & B \\ C & D \end{pmatrix} \in \mathrm{Sp}(m,\mathbf{Z})$$

ith the special properties

$$\left\{\begin{matrix} A = I, & B = \begin{pmatrix} 0 & g \\ g^t & 0 \end{pmatrix} \\ C = 0, & D = I \end{matrix}\right\} \quad \text{and} \quad \left\{\begin{matrix} A = \begin{pmatrix} I_1 & 0 \\ g^t & 1 \end{pmatrix}, & B = 0 \\ C = 0, & D = \begin{pmatrix} I_1 & -g \\ 0 & 1 \end{pmatrix} \end{matrix}\right\}$$

here B and A, D are partitioned in the same way as Z, g is a $(n-1)$-ector with elements in $\mathbf{Z}$, and I the unit matrix. These transformations eave the Fourier expansion untouched and show that

$$\phi_0(Z_1,z) = \phi_0(Z_1,z+g) = \phi_0(Z_1,z+Z_1g).$$

o ϕ_0 is $2n$-fold periodic with the period matrix (I_1,Z_1). ϕ_0 is also olomorphic. A holomorphic function of this sort must be a constant. o ϕ_0 does not depend on z.

Further we apply the transformations

$$M = \left\{ \begin{array}{ll} A = \begin{pmatrix} A_1 & 0 \\ 0 & 1 \end{pmatrix}, & B = \begin{pmatrix} B_1 & 0 \\ 0 & 0 \end{pmatrix} \\ C = \begin{pmatrix} C_1 & 0 \\ 0 & 0 \end{pmatrix} & D = \begin{pmatrix} D_1 & 0 \\ 0 & 1 \end{pmatrix} \end{array} \right\}, \quad M_1 = \begin{pmatrix} A_1 & B_1 \\ C_1 & D_1 \end{pmatrix} \in \mathrm{Sp}(m-1,\mathbf{Z})$$

and see that $\phi_0(Z_1)$ is a Siegel modular form in Z_1 with the same weight as $f(Z)$. One usually writes

$$\Phi f(Z) = \phi_0(Z_1).$$

The operator Φ operates of course as well on modular forms for a subgroup of $\mathrm{Sp}(m,\mathbf{Z})$.

The $(m-1)$-fold application of Φ maps a Siegel modular form on a classical modular form of the same weight.

Theorem. Φ^{m-1} *maps the domain* $J^{(4)}$ *of all Siegel modular forms of weights which are divisible by* 4 *on the domain* $J_1^{(4)}$ *of classical modular forms which is generated by* $g_2(z)$ *and* $g_6(z)$ *where* $g_2(z)$ *has been defined above and* $g_6(z)$ *is a modular form on* z *of weight* 12 *which is independent of* $g_2(z)$.

For the proof we need a short preparation. Clearly we must explicitly point out those Siegel modular forms $f_4(Z)$ and $f_{12}(Z)$ which are mapped on $g_2(z)$ and $g_6(z)$. There are chiefly two methods of construction, the Eisenstein series and their generalizations (which we mentioned in the Hilbert case), and the theta series. The latter are much easier to handle in many respects, and they are of more immediate interest to the number theorist. The proof that the theta series are Siegel modular forms, as found in literature, can be simplified.

We consider the matrix F of a definite quadratic form in l variables. The elements of F are assumed as rational integers, the elements in the diagonal as even. The least positive integer L for which LF^{-1} has again such coefficients is called the *level* of F. What we want to show is that the theta series

$$\theta_F(Z) = \sum_X e^{\pi i \, tr(X^t F X Z)} \tag{2}$$

to be summed over all matrices X with coefficients in $\mathbf{Z}$ of l rows and n columns is a Siegel modular form with respect to the congruence subgroup

$$\Gamma_0(L): \begin{pmatrix} A & B \\ C & D \end{pmatrix} \equiv \begin{pmatrix} * & * \\ 0 & * \end{pmatrix} \bmod L.$$

We can write (2) in a different way by introducing the ml-square matrix (the Kronecker product)

$$\mathcal{Z} = Z \times F = \begin{pmatrix} z_{11}F & \cdots & z_{1n}F \\ & \vdots & \\ z_{n1}F & \cdots & z_{nn}F \end{pmatrix}.$$

Namely now (2) is

$$\theta_F(Z) = \Theta(\mathcal{Z}) = \sum_x e^{\pi i x^t \mathcal{Z} x} \tag{3}$$

where x runs over all ml-vectors with elements in $\mathbf{Z}$.

Now the following substitutions are identical:

$$Z \longrightarrow (AZ+B)(CZ+D)^{-1} = M(Z) \quad \text{and} \quad \mathcal{Z} = (\mathcal{A}\mathcal{Z}+\mathcal{B})(\mathcal{C}\mathcal{Z}+\mathcal{D})^{-1} = \mathcal{M}(\mathcal{Z})$$

with

$$\mathcal{M} = \begin{pmatrix} \mathcal{A} = A \times I & \mathcal{B} = B \times F \\ \mathcal{C} = C \times F^{-1} & \mathcal{D} = D \times I \end{pmatrix}$$

I the identity matrix. It is known (for instance: M. Eichler, Introduction to the Theory of Algebraic Numbers and Functions, Appendix to Chapter I, §1) that

$$\Theta(\mathcal{M}(\mathcal{Z}))|\mathcal{C}\mathcal{Z}+\mathcal{D}|^{-\frac{1}{2}} = \chi(\mathcal{M})\Theta(\mathcal{Z}) \tag{4}$$

with a certain 8-th root of unity

$$\chi(M) = \chi(\mathcal{M})$$

if $\mathcal{M} = \begin{pmatrix} A & B \\ C & D \end{pmatrix}$ is contained in the subgroup $\Theta \subset Sp(ml,\mathbf{Z})$ defined by the property that C^tA and B^tD are matrices with even elements in the diagonal. Θ is a subgroup of finite index in $Sp(ml,\mathbf{Z})$. The root of unity has the properties $\chi(M) = \chi(\mathcal{M}) = 1$ for $M = \begin{pmatrix} I & S \\ 0 & I \end{pmatrix}$ and $M = \begin{pmatrix} 0 & I \\ -I & 0 \end{pmatrix}$ if S is any integral symmetric matrix. The first fact is obvious from (2). The second follows by inserting $Z = iE$ and applying this M which yields

$$\Theta(iI \times F^{-1})|iF|^{-\frac{m}{2}} = \Theta(iI \times F).$$

By a theorem of Witt (cf. the above quoted book), $\begin{pmatrix} I & S \\ 0 & I \end{pmatrix}$ and $\begin{pmatrix} 0 & I \\ -I & 0 \end{pmatrix}$ generate the whole group $Sp(m,\mathbf{Z})$. From the definition of L and $\mathcal{M}$ we see that $\mathcal{M}$ lies in the subgroup Θ whenever M lies in the subgroup $\Gamma_0(L)$. So (4) yields

$$\theta_F(M(Z))|CZ+D|^{-\frac{l}{2}} = \theta_F(Z) \quad \text{for} \quad M \in \Gamma_0(L). \tag{5}$$

Quadratic forms of level $L = 1$ exist if the number of variables is $l \equiv 0 \bmod 8$. So we obtain at first a Siegel modular form of weight 4 for the whole group $\Gamma_0(1) = \Gamma = Sp(m,\mathbf{Z})$ which we shall denote by $f_4(Z)$. Applying the operator Φ^{m-1} to its Fourier series we are led to

$$\Phi^{m-1}f_4(Z) = \phi_4(z) = \sum_x e^{\pi i z x^t F_8 x}, \tag{6}$$

summed over all 8-vectors x with coefficients in $\mathbf{Z}$. This is a classical modular form of weight 4. It is known that all such modular forms are constant multiples of the elementary Eisenstein series $g_2(z)$. Similarly we take a quadratic form F_{24} in 24 variables and form the corresponding series (3) which we call $f_{12}(Z)$. It is mapped by Φ^{m-1} on

$$\Phi^{m-1}f_{12}(Z) = \theta_{12}(z) = \sum_x e^{\pi i z x^t F_{24} x}. \tag{7}$$

One can show that a quadratic form F_{24} of level 1 exists for which $_{12}(z)$ is independent of $\theta_4(z)$. But the proof involves deep tools from umber theory. Consider all quadratic forms F_{24} of this nature. iegel's theory states that a certain linear combination of their theta eries is equal to the Eisenstein series $g_6(z)$. The Fourier expansions f $g_2(z)$ and $g_6(z)$ are explicitly known, and one can check that $g_6(z)$ s not a constant multiple of $g_2(z)^3$. So there must exist at least a $_{24}$ whose theta series $\theta_{12}(z)$ is linearly independent of $\theta_4(z)$ and ence independent at all. With this the theorem is proved.

18. *Principal ideals*

The theory of Abelian functions can be built up similarly to the heory of modular functions. One starts with the theta functions. hese can be assigned "weights," and with these they form a finitely enerated graded ring. The quotients of theta functions of equal eights are the Abelian functions. Thus the theta functions provide a rojective model for the Abelian variety. An important theorem of *ppell* and *Humbert* states that a prime divisor in the sense of this model is given by the variety of zeros of one particular theta function. ee for instance *F. Conforto*: Abelsche Funktionen und algebraische eometrie, Springer-Verlag, Berlin-Göttingen-Heidelberg 1956, p. 184. e want to prove an analogous theorem for Hilbert modular forms, $n = 2$. e make two further assumptions.

a) *by the specialization* $z^1 = z^2 = z$ *the ring* J *of Hilbert modular forms is mapped on a ring* J_1 *of elliptic modular forms in* z*, in hich all reflexive ideals are principal (RPID).*

A sufficient condition for this has been discussed in §17, but here may be other cases.

With the kernel of this map P we have an exact sequence

$$\cdots \longrightarrow \mathrm{Ext}^1_h(J,h) \longrightarrow \mathrm{Ext}^1_h(P,h) \longrightarrow \eta^{-1}\mathrm{Ext}^1_{h_1}(J_1,h_1) = 0.$$

By the hypothesis of §16 and the assumption on J_1, implying that J_1 is reflexive, we find that $\mathrm{Ext}^1_h(P,h) = 0$, and P is a free module.

b) *the intersection number* $d(M,P)$ *of* P *with every reflexive ideal* $M \subset J$ *(which is not divisible by* P*) is positive.*

Theorem 1. *A reflexive J-ideal* $M \subset J$ *which is not divisible by* P *is a principal ideal if and only if it is quasiinvertible, and if* $M \cap P$ *is a free h-module.*

Theorem 2. *Now we suppose only the specialization* $z^1 = z^2$ *to satisfy condition* a). *An J-ideal* $M \subseteq J$ *which is not divisible by* P *is a principal ideal if and only if it is quasiinvertible, and* $M \cap P$ *and,* M *chosen in such a way that* $MM^{-1} \subset J$ *is prime to* P, $MM^{-1} \cap P$ *are free h-modules.*

Proofs. That the conditions are necessary follows from the hypothesis of §16.

If $M \cap P$ is free, $M_1 = M/(M \cap P)$ is reflexive by Theorem 1 in §8 and hence a principal ideal $J_1 M_1$. Let M be an element in M which is mapped on M_1. Then $JM \subseteq M$. If the equality sign would not hold, $MM^{-1} = U$ would be an integral ideal, not J, and not divisible by P. Now we have $MU \approx JM$, and by Theorem 1 in §14

$$d(JM,P) = d(M,P) + d(U,P).$$

Furthermore by §14

$$d(JM,P) = d(JM/(JM \cap P)) = d(J_1M_1),$$

$$d(M,P) = d(M/(M \cap P)) = d(J_1M_1).$$

This is impossible under the assumption b) on P, and Theorem 1 is roved.

Under the assumptions of Theorem 2 we have

$$U/(U \cap P) = J_1 U_1$$

nd

$$J_1 M_1 = JM/(JM \cap P) \approx MU/(MU \cap P) = (M/(M \cap P))(U/U \cap P))$$

$$= J_1 M_1 U_1 U_1 .$$

ecause the modules at the ends are reflexive, they are equal, and herefore U_1 is a constant. By the second isomorphy theorem

$$0 = J_1/U_1 = (J/P)/(U/(U \cap P)) = J/(U+P).$$

f $U \neq J$, $U + P$ would not contain 1, which would contradict this quation. So we have $M^{-1} = J$. Multiplying by M and using the quasi-nvertibility we get

$$JM \approx MM^{-1}M = M.$$

ecause both modules are reflexive, they are equal. This proves heorem 2.

Theorem 3. *Under the assumptions* a) *and* b), *and if* P *is quasi-nvertible,* P *is a principal ideal.*

Proof. Because of the quasiinvertibility we can find $\bar{P} \subseteq J$, ot divisible by P, for which $P \cap \bar{P}$ is principal, Then by Theorem 1 $\bar{P}$ is principal and consequently also P.

Corollary to Theorem 3. *Under the conditions of Theorem* 3, M *is rincipal if and only if* M *is quasiinvertible and a free* h*-module.*

Indeed the freeness of $M \cap P = MP$ follows from that of M.

Remark. If $P = JP$ is principal and maps J on the ring $J_1 = \mathbf{C}[g_2, g_3^2]$, J is generated by the Eisenstein series $G_2(z^1, z^2)$, $G_3(z^1, z^2)$, and P.

This is obvious. In §19 we shall see that the consequence is wrong in most cases. Thus the meaning of Theorem 3 can only be that the ideal P cannot have in general such simple properties. There is one case, however, when J is generated by G_2, G_3, P, namely when $d = 3$ and J is the subring of "symmetric" modular forms (see §19, No. 5).

§19. *Hilbert modular forms in 2 variables*

1. *Introduction.*

In this last section we compute the intersection numbers of certain prime divisors. They are estimated by sums over class numbers of imaginary quadratic number fields. They will be equal to the intersection numbers studied in §14, if the intersection points are simple and not singular. If a prime divisor avoids all singular points at all, its ideal of multiples is quasiinvertible (§15), and we can apply the theory of §14. In all cases when the genus coefficients of the occurring ideals are known we arrive at inequalities for class numbers of many imaginary quadratic fields which, because they baffle intuition, are an excuse for long and tedious work.

$\Omega = \mathbf{Q}(\sqrt{d})$ with square-free d is a real quadratic number field and o its principal order. The non-identical automorphism will be written $\alpha \longrightarrow \alpha'$. The two complex variables of the Hilbert modular forms are now z, z' (instead of z^1, z^2). We shall be using an admissible coordinate ring h, of weight h, and H will denote the ring of al modular forms of weights divisible by h.

We will study two specializations. The first is $z = z'$, mapping H on a subring H_1 of all classical modular forms. The kernel will be written P_0:

(1) $$H/P_0 = H_0.$$

The second specialization is as follows: let q be a positive rational integer. By the substitution

$$f(z,z') \longrightarrow \phi(z) = f(z, -\frac{q}{z})z^{-h}$$

a modular form $f(z,z')$ of weight l is mapped on an automorphic form $\phi(z)$ of weight $2l$ with respect to the group

(2) $$\Gamma_q = \left\{ \begin{pmatrix} \alpha & -q\beta \\ \beta' & \alpha' \end{pmatrix} \in \Gamma \right\}.$$

The kernel is another prime ideal P_q:

(3) $$H/P_q = H_q.$$

We shall sometimes briefly speak of the "curves" P_0 or P_q, meaning the curves of their zeros. The automorphic forms $\phi(z)$ have been discovered by *Poincaré* and extensively studied by *G. Shimura* (On the theory of automorphic functions, Annals of Math. 70(1959),101-144).

2. *How to avoid singular points.*

We have just given reasons why the curves P_q should avoid singularities. We now seek conditions for that.

The group (2) is a representation of the unit group of an order in the quaternion algebra Φ_q/Q defined by the quadratic form $x_0^2 - dx_1^2 - qx_2^2 - dqx_3^2$. The first assumption which we make on q is that Φ_q has no divisors of zero. Then the group (2), as a group of fractional linear substitutions of the upper half-plane, has a compact fundamental domain. In this case the curve of zeros of P_q does not meet the "cusps" which are singularities of the modular variety.

The other singularities are the fixed points of elements $M \neq I \in \Gamma$ of finite order. Let z, z' be such a point, lying on P_q. Then we have the equations

$$(4) \qquad M(z) = \frac{\alpha z+\beta}{\gamma z+\delta} = z, \quad M'(z') = z', \quad zz' = -q.$$

Using the abbreviation

$$(5) \qquad J_q = \begin{pmatrix} 0 & -q \\ 1 & 0 \end{pmatrix}.$$

We can write (4) thus:

$$M(z) = z, \quad J_q^{-1}M'J_q(z) = z.$$

Both equations have a common solution if and only if

$$(6) \qquad J_q^{-1}M'J_q = uI + vM$$

with real u and v. Inserting the coefficients of $M = \begin{pmatrix} \alpha & \beta \\ \gamma & \delta \end{pmatrix}$ we find

$$\gamma' = -\frac{v}{q}\beta, \quad v\gamma = -\frac{1}{q}\beta'.$$

So either is $v = 0$ and $\gamma = \beta = 0$, but then M has not a fixed point in the upper half plane. Or $v = \pm 1$. In this case the trace of (6) shows

$$\alpha' + \delta' = 2u \pm (\alpha+\delta).$$

If $v = 1$, we have $u = 0$, and M has the form $M = \begin{pmatrix} \alpha & -q\beta \\ \beta' & \alpha' \end{pmatrix}$ (we have written $-q\beta$ instead of β). So M belongs to the group (2). We can prevent this by demanding that Φ_q contains no element of finite order except ± 1. The conditions for this are known (see the following proposition, part 2).

It remains $v = -1$. The possible finite orders of elements $M \neq \pm I \in \Gamma$ are 3, 4, 5, 6, 8, 10. Orders 5, 8, 10 occur only for $d = 2$ or 5. We shall exclude these later for other reasons. For $M^4 = I$ we have $\alpha + \delta = \alpha' + \delta' = 0$ and $u = 0$, because of (6):

$$M = \begin{pmatrix} \alpha & q\gamma' \\ \gamma & -\alpha' \end{pmatrix}, \qquad \alpha - \alpha' = 0.$$

So α is rational and then

$$1 = |M| = -\alpha - q\gamma\gamma'.$$

t follows that -1 is a quadratic residue for all odd primes dividing . For $M^3 = I$ we have $\alpha + \delta = \alpha' + \delta' = -1$ and $u = -1$, because of (6)

$$M = \begin{pmatrix} \alpha & q\gamma' \\ \gamma & -\alpha'-1 \end{pmatrix}, \qquad \alpha - \alpha' - 1 = -1.$$

So α is again rational and then

$$1 = |M| = -\alpha^2 - \alpha - q\gamma\gamma'.$$

t follows that -3 is a quadratic residue for all odd primes dividing .

Gathering all conditions on q together we have

Proposition 1. *Let* $d \neq 2$ *and* $\neq 5$. *On* P_q *lies no singular oint if and only if the following conditions are satisfied:*

1) *The quaternion algebra* Φ_q, *represented by the matrices* $\begin{pmatrix} \alpha & -q\beta \\ \beta' & \alpha' \end{pmatrix}$ *does not contain divisors of zero.*
2) Φ_q *does not contain a quadratic subfield of discriminant* -3 *or* -4.
3) *There exists an odd prime divisor* q_1 *of* q *with* $\left(\frac{-1}{q_1}\right) = -1$.
4) *There exists an odd prime divisor* q_2 *of* q *with* $\left(\frac{-3}{q_2}\right) = -1$. *It is not excluded that* $q_1 = q_2$. *Instead of* 3) *or* 4) q *may be divisible by* 4.

3. *The number of common points of P_0 and P_q.*

We can calculate the algebraic intersection number $d(P_0, P_q)$ if one of the ideals is principal. But then we need the weight of the form generating this ideal.

Its determination requires number theoretical considerations in the algebra Φ_q. We begin with some preparations. The matrices $\begin{pmatrix} \alpha & -q\beta \\ \beta' & \alpha' \end{pmatrix}$ with $\alpha, \beta \in o$ form an order O of Φ_q of discriminant $(4dq)^2$. The only orders for which ready methods and results are available are those with discriminants equal to a square of a square-free number, they have been called *orders of square-free level* (see M. Eichler, Zur Zahlentheorie der Quaternionen-Algebren, Journ. reine angew. Math. **195**(1956), 127-151). So we assume from now on d and q as odd and consider the larger order (this time the prime does not indicate the automorphism of Ω) defined by

$$(7)\quad O': M = \frac{1}{2}\begin{pmatrix} \alpha & -q\beta \\ \beta' & \alpha' \end{pmatrix}, \qquad \begin{array}{l} \alpha = a_1 + a_2\sqrt{d}\,, \quad \beta = b_1 + b_2\sqrt{d}\,, \\ a_1 \equiv a_2 \equiv b_1 \equiv b_2 \bmod 2 \quad \text{and} \in Z. \end{array}$$

That O' is indeed an order one sees by checking with

$$\alpha = \pm 1 \pm \sqrt{d} + \alpha_0, \qquad \beta = \pm 1 \pm \sqrt{d} + \beta_0, \qquad \alpha_0, \beta_0 \in 4o,$$

if not $\alpha, \beta \in 2o$. The level (square root of the discriminant) is $2dq$, and square-free if d, q are odd, square-free and relatively prime. We shall use the splitting up

$$(8)\qquad 2dq = l_1 l_2$$

where l_1 is the product of the prime numbers ramified in Φ_q.

The maximal commutative suborders $i \in O'$ are 2-dimensional $\mathbf{Z}$-modules $i = \mathbf{Z}(1, \omega)$. If $\Delta(i) = \mathrm{discr}(1, \omega)$ is their discriminant, they are orders in commutative quadratic fields $\mathbf{Q}(\sqrt{\Delta(i)})$.

Proposition 2. A rational integer Δ occurs as the discriminant $\Delta = \Delta(i)$ of a maximal suborder $i \subset O$ if and only if the modified *egendre symbol* $\left\{\frac{\Delta}{p_1}\right\} \neq 1$ *for all primes* p_1 *dividing* l_1 *and* $\left\{\frac{\Delta}{p_2}\right\} \neq$ 1 *for all primes* p_2 *dividing* l_2.

The proof as well as the explanation of the modified Legendre symbol are to be found in the paper quoted above. We need yet the index of the groups Γ_q and $\Gamma_q{}'$ of units of norm 1 of O and O':

$$[\Gamma_q':\Gamma_q] = \begin{cases} 3 & \text{for } d \equiv -q \equiv 3 \bmod 4, \\ 1 & \text{in all other cases.} \end{cases} \tag{9}$$

or the proof we form the determinant

$$|M| = \frac{1}{4}(a_1^2 - da_2^2 + qb_1^2 - qdb_2^2).$$

his number is even for odd a_i, b_i, except for $d \equiv -q \equiv 3 \bmod 4$, when t is odd. So in the upper cases of (9) there exist no units with a_i, b_i dd, and then all elements of $\Gamma_q{}'$ and Γ_q are the same. In the exceptional case Φ_q is ramified at 2, and the 2-adic extension O_2' is a maximal order. The elements of O are characterized in O by their traces eing even. Let Z_2' be the prime O'-ideal of norm 2. Then O'/Z_2' is commutative field, namely the field of the 3rd roots of unity mod 2. $_q$ being an indefinite quaternion algebra contains in each maximal rder units M with odd traces (M. Eichler, Allgemeine Kongruenzklasseneinteilungen der Ideale einfacher Algebren und ihre L-Reihen, Journ. eine angew. Math. **179**(1938), 227-251, Hilfssatz 5). Such a unit M is 3rd root of unity. Therefore $M^3 \in O$. This proves (9).

After this preparation we will study the common points of P_0 and $_q$. These curves are given by

$$P_0:\ M_1'(z') = M_1(z),$$

$$P_q:\ M_2'(z') = J_q M_2(z),$$

ith $M_1, M_2 \in \Gamma$. J_q had been explained by (5). Without loss of generality we may take $M_2 = I$, and we write $M = \begin{pmatrix} \alpha & \beta \\ \gamma & \delta \end{pmatrix}$ for M_1. By elimination of z' we obtain $M' J_q(z) = M(z)$. Abbreviating this complex number by ζ we have to solve

(10) $$M' J_q M^{-1}(\zeta) = \zeta$$

by a ζ in the upper half plane.

The matrix

(11) $$P = M' J_q M = \begin{pmatrix} q\alpha'\gamma + \beta'\delta & -q\alpha\alpha' - \beta\beta' \\ q\gamma'\gamma + \delta'\delta & -q\alpha\gamma' - \beta\delta' \end{pmatrix}$$

satisfies the equation

(12) $$PP' = -qI.$$

We write

(13) $$P = P_1 + \sqrt{d}\, P_2$$

where P_1, P_2 have coefficients in $\mathbf{Z}$. The necessary and sufficient condition for (10) to have a solution in the upper half-plane is the inequality

(14) $$|\mathrm{tr}(P)| < 2\sqrt{q} = 2\sqrt{|P|}\ .$$

From (12) and (13) follows

(20) $$P_1^2 - dP_2^2 = -qI, \qquad P_1P_2 - P_2P_1 = 0.$$

P_1 is not a multiple of I, because of (11) and $q\alpha\alpha' + \beta\beta' \neq 0$, the commutativity of P_1 and P_2 shows that $P_2 = uI + vP_1$ with rational u, v. Apparently from (11) $v = 0$, and now (20) yields

(21) $$P_1^2 = (du^2 - q)I,$$

and (19) becomes

$$|u| < \sqrt{\frac{q}{d}} \; . \tag{22}$$

n the converse direction, we do not know whether every solution of 21) has the form (11) with an $M \in \Gamma$.

If we transform P_1 by a matrix M_0 from the classical modular group L(2,**Z**), M is taken into $M_0 M$ and ζ into $M_0(\zeta)$. So the number of common points of P_0 and P_q is equal or smaller than the number of classes of equivalent matrices satisfying (21) and (22). This number is

$$d(P_0,P_q) \leq \sum_{u,f} h(4(du^2-q)f^{-2}), \tag{23}$$

o be summed over all integral u with (22) and all positive integral f uch that $\Delta = 4(du^2 - q)f^{-2}$ is the discriminant of an order in an maginary quadratic field. $h(\Delta)$ is the class number of ideals of this rder. We have already denoted this number by $d(P_0,P_q)$ anticipating hat it is equal to the intersection number defined in §14.

The number of common points of two P_q.

Now we calculate the number of common points of two such ideals P_{q_1}, P_{q_2}. The procedure is similar as in the previous section. We ake the assumptions

$$d \equiv 7 \bmod 8, \quad q_1 \equiv 3 \bmod 8, \quad q_2 \equiv 0 \bmod 2, \tag{24}$$

nd also that q_1 is square-free and relatively prime to d. The common oints are common solutions of

$$P_{q_1} : M_1'(z') = J_{q_1} M_1(z),$$

$$P_{q_2} : M_2'(z') = J_{q_2} M_2(z).$$

e may again put $M_1 = M = \begin{pmatrix} \alpha & \beta \\ \gamma & \delta \end{pmatrix}$, $M_2 = I$. This gives us

$$M' J_{q_2} M^{-1} J_{q_1}^{-1}(\zeta) = \zeta$$

ith a $\zeta = z$ in the upper half-plane. As a fractional linear transformation $J_{q_1}^{-1}$ and $-J_{q_1}$ yield the same. We abbreviate

$$(25) \qquad P = -M' J_{q_2} M^{-1} J_{q_1} = \begin{pmatrix} q\alpha\alpha' + \beta\beta' & q_1 q_2 \alpha'\gamma + q_1 \beta'\delta \\ q_1 \alpha\gamma' + \beta\delta' & q_1 q_2 \gamma\gamma' + q_1 \delta\delta' \end{pmatrix}$$

and

$$(26) \qquad P = \frac{1}{2}\left(uI + \frac{1}{\sqrt{d}} P_2\right), \qquad P_2 = \begin{pmatrix} v\sqrt{d} & -2q_1\beta \\ 2\beta' & -v\sqrt{d} \end{pmatrix}.$$

P_2 has trace 0, and therefore

$$(27) \qquad P_2^2 = (u^2 - 4q_1q_2)dI .$$

The condition that P has a fixed point in the upper half-plane is

$$(28) \qquad |u| < 2\sqrt{q_1q_2} .$$

P_2 belongs to the order $\mathcal{O}$, with $q = q_1$. Conversely, a $P_2 \in \mathcal{O}$ satisfying (27), has the form

$$P_2 = \frac{1}{2}\begin{pmatrix} \alpha_1 & -q_1\beta_1 \\ \beta_1' & \alpha_1' \end{pmatrix}, \qquad \alpha_1 + \alpha_1' = 0,$$

or

$$P_2 = \begin{pmatrix} v\sqrt{d} & -q_1\beta_2 \\ \beta_2' & -v\sqrt{d} \end{pmatrix} \quad \text{with} \quad v \in \mathbf{Z}, \ \beta_2 \in \mathfrak{o} .$$

From (27) follows also

$$dv^2 q_1 \beta_2 \beta_2' = (u^2 - 4q_1q_2)d ,$$

and if $\beta_2 = (x+y\sqrt{d})\sqrt{d}$ with $x,y \in \mathbf{Z}$, under the assumption (24),

$$v^2 + 5x^2 + 5y^2 \equiv u^2 \bmod 8.$$

This is only possible for $v \equiv u \bmod 2$ and $x^2 + y^2 \equiv \beta_2\beta_2' \equiv 0 \bmod 4$. Because $d \equiv 7 \bmod 8$, $\beta_2 \equiv 0 \bmod 2$, and so P_2 has the form given in (26) and belongs even to $\mathcal{O}$.

We are interested in the classes $M_0^{-1}P_2M_0$ with $M_0 \in \Gamma_q$ of matrices P_2 satisfying (27), (28). The number of these classes is equal to the number of classes $M_0'^{-1}P_2M_0'$ with $M_0' \in \Gamma'_{q_1}$, but multiplied by the index (9). (This time the prime does not indicate the automorphism of Ω.) According to the assumptions (24), this index is 1. The number of the latter classes is less than or equal to

$$(29) \quad d(P_{q_1},P_{q_2}) \leq \sum_{u,f} \prod_{p_1/l_1} \left(1 - \left\{ \frac{(u^2-4q_1q_2)f^{-2}}{p_1} \right\} \right) \prod_{p_2/l_2} \left(1 + \left\{ \frac{(u^2-4q_1q_2)f^{-2}}{p_2} \right\} \right) h((u^2-4q_1q_2)f^{-2})$$

to be summed over all integral u with (28) and all positive integral f such that

$$\Delta = (u^2-4q_1q_2)f^{-2}$$

is the discriminant of an order in an imaginary quadratic field. We cannot produce the proof here in detail and only refer to the paper quoted on page 104. According to proposition 2 the product after the sum sign is 0 whenever there does not exist a suborder $i \in O'$ with discriminant Δ, and it is equal to the index of the group of rational ideals in the group of ambiguous ideals for such an order i. The proof of (29) is based on a theorem of Chevalley, Hasse, and E. Noether on ideals in hypercomplex systems.

. *The curve* $z = z'$.

We use slightly different notations than in §18. J means the ring of all modular forms of even weights, Q_0 the kernel of the specialization $z = z'$, and

$$(30) \qquad J/Q_0 = J_0 = \mathbf{C}[g_2, g_3^2].$$

The intersection with H is our former ideal

$$(31) \qquad P_0 = H \cap Q_0 \quad \text{and} \quad H/P_0 = H_0.$$

Let us assume (cf. §18, Theorem 3) that

(32) $$Q_0 = JP.$$

Proposition 3. *Under the hypothesis of* §16 *and the assumption that* Q_0 *intersects with every positive divisor, the weight of* P *is*

(33) $$w = \frac{2\pi^4 h(\sqrt{d})}{3D(\sqrt{d})^{3/2}\zeta(2,\sqrt{d})},$$

where $D(\sqrt{d})$, $h(\sqrt{d})$, $\zeta(s,\sqrt{d})$ *are the discriminant, the ideal class number, and the zeta function of* $\Omega = \mathbf{Q}(\sqrt{d})$.

Proof. The image of a form $f \in J$ of weight λh under the specialization is a classical modular form f_1 of weight $2\lambda h$ in $J_0 = \mathbf{C}[g_2, g_3^2]$ and therefore a polynomial

$$f_1 = \gamma_0 g_2^{\frac{\lambda h}{2}} + \gamma_1 g_2^{\frac{\lambda h}{2}-3} g_3^2 + \cdots + \gamma_r g_3^{\frac{\lambda h}{3}}.$$

The number of linearly independent f_1 of weights $2\lambda h$ is $L(2\lambda h) = \frac{\lambda h}{6} + 1$. On the other hand, this number is expressed by the rank polynomial which is $N_1\lambda$ + a constant, where N_1 is the degree of the algebraic extension of H_0 over h. By comparison we get

$$N_1 = \frac{h}{6}.$$

Calculating the rank polynomial by §6, (5) we get

(34) $$G(P_0) = G(H) + \frac{h}{6}.$$

If we replace P_0 by its h-th power, which is HP^h, the difference of the linear degrees is multiplied by h. P^h is a modular form of weight wh; but as an algebraic form in the y_ν, its degree is only w. On the other hand, the linear degree of HP^h is $G(H) + wN$, where N is the degree of the extension H over h. This leads to

(35) $$w = \frac{h^2}{6N}.$$

The degree N is obtained by comparing the rank polynomial

$$H(\lambda,H) = N\binom{\lambda}{2} + \gamma_1\lambda + \gamma_0$$

for the number of linearly independent elements of degree λ with the rank formula of *Shimizu* (On discontinuous groups operating on the product of upper half-planes, Annals of Math. **77**(1963), 33-71) which gives the number of linearly independent modular forms of weight $2r = \lambda h$:

$$H(\lambda,H) = \frac{D(\sqrt{d})^{3/2}\zeta(2,\sqrt{d})}{8\pi^4 h(\sqrt{d})}(2r-1)^2 + \cdots$$

Equating the highest terms we get

$$\frac{N}{h^2} = \frac{D(\sqrt{d})^{3/2}\zeta(2,\sqrt{d})}{4\pi^4 h(\sqrt{d})} \tag{36}$$

and lastly (33).

Example. For $d = 3$ we get from (33) the weight of the form P generating $P_0 = JP$ as $w = 1$ which is impossible since all forms in J are assumed of even weights. So our assumptions are wrong.

Gundlach (quotation see §16) has considered the larger group $\hat{\Gamma}_\varepsilon$ which is generated by the adjunction of the substitutions

$$\begin{matrix} z \longrightarrow \varepsilon z & \quad\text{and}\quad & z \longrightarrow z' \\ z' \longrightarrow \varepsilon' z' & & z' \longrightarrow z \end{matrix}$$

where ε is the basic unit of Ω. The index of Γ in $\hat{\Gamma}_\varepsilon$ is 4. Our theory applies also for $\hat{\Gamma}_\varepsilon$, and N has to be divided by 4. This gives us $w = 4$. Indeed Gundlach has shown that $P = G_4 - G_2^2$. In this case all our assumptions are correct.

It would even suffice to replace Γ by Γ_ε, formed by the first adjunction only. Then $w = 2$, and we see that $P = \sqrt{G_4 - G_2^2}$. This modular form is invariant under Γ_ε, but changes the sign under $z \longleftrightarrow z'$.

In the next step $\Gamma_\varepsilon \longrightarrow \Gamma$ again a square root has to be adjoined, but now $\sqrt{P}$ becomes divisible by two different prime divisors, one of which is P_0. J remains a free k-module, but P_0 does now not intersect with every positive divisor. So our variety already differs material from a projective plane.

6. *The algebraic intersection number of* P_0 *and* P_q.

Maintaining our assumptions of the last section we now calculate the intersection number $d(P_0,P_q)$ for a P_q which avoids singular point According to Theorem 1 in §14 we have

$$d(P_0,P_q) = \frac{1}{h}\, d(P_0^h,P_q) = \frac{1}{h}\, d(HP^h,P_q)$$

$$= \frac{1}{h}\, (\gamma(H)-\gamma(HP^h)-\gamma(P_q)+\gamma(P_qP^h)).$$

On the other hand, for any module M, and any element $f \in H$ of degre w

$$\gamma(Mf) = N(\frac{w(w+1)}{2} - 2w) + G(M)w + \gamma(M).$$

With this we get

$$(36) \qquad d(P_0,P_q) = \frac{w}{h}\, (G(P_q)-G(H)) = \frac{h}{6N}\, (G(P_q)-G(H)).$$

The second factor on the right can be computed as follows. P_q maps a modular form of weight λh on an automorphic form of weight $2\lambda h$ The coordinates y_0, y_1 double their weights, too. Let K_q be the fiel of these automorphic functions and $k = \mathbf{C}(x)$, $x = y_1 y_0^{-1}$. The degree is

$$(37) \qquad G(P_q) - G(H) = [K_q : k],$$

by the same argument already used in No. 5.

Our specialization maps a modular form $f(z,z')$ of weight k on $f(z,z')z^{-k} = \phi(z)$. On the other hand, such a modular form represents the differential of degree k: $f(z,z')(dz\, dz')^k$, and this is mapped on the differential $\phi(z)\, dz^{2k}$. Especially the $y_\nu(dz\, dz')$ are mapped n differentials $y_\nu dz^{2h}$. They are of course integral.

The number of linearly independent integral differentials of de-ree $2\lambda h$ is given by the theorem of Riemann-Roch as $(g-1)(4\lambda h-1)$, where is the genus of K_q. Then same number can be given by the rank poly-omial as $[K_q:k]\lambda$ + a constant. This implies

$$[K_q:k] = 2h(g-1). \tag{38}$$

The genus is given by the formula

$$g-1 = \frac{\phi(2dq)}{12}[\Gamma'_q:\Gamma_q] = \frac{\phi(dq)}{12}[\Gamma'_q:\Gamma_q] \tag{39}$$

here ϕ is the Euler function and Γ'_q is the full group of units of orm 1 in a maximal order containing O (*M. Eichler*, Über die Einheiten er Divisionsalgebren, Math. Annalen **114**(1937), 635-654). The order onsidered in No. 3 is the maximal order, if d is odd and square free, nd q the product of an even number of primes not dividing $2d$. The ndex $[\Gamma'_q:\Gamma_q]$ can now be taken from (9).

Gathering all results together we arrive at

Proposition 4. *Under the following conditions the intersection umber of* P_0 *and* P_q *in the variety attached to the group* Γ_q *is*

$$d(P_0,P_q) = \frac{2\pi^4 h(\sqrt{d})}{9D(\sqrt{d})^{3/2}\zeta(2,\sqrt{d})}\phi(dq)[\Gamma'_q:\Gamma_q] \tag{41}$$

here ϕ *is the Euler function and the index* $[\Gamma'_q:\Gamma_q]$ *is given in* (9).

Conditions:

1) P_q *avoids all singular points (cf. Proposition* 1*).*
2) d *is odd and square-free, and* q *is the product of an even number of primes not dividing* $2d$*.*
3) *The assumptions of No.* 5 *on the curve (for instance* $d = 3$*).*

Since all intersections are simple, $d(P_0,P_q)$ *coincides with the umber studied in No.* 4 *and is therefore less than or equal to the sum f class numbers given in* (23).

7. *The intersection number of P_{q_1} and P_{q_2}.*

It may be that the kernels of the maps $z' = -\frac{q}{z}$ are principal J-ideals (see §18, Theorem 1):

$$Q_q = JP_q \ .$$

Then their h-th powers are also principal: $P_q^h = HP_q^h$. If the weights of the P_q are called w_q, the intersection number is

$$d(P_{q_1},P_{q_2}) = \frac{1}{h^2}\, d(HP_{q_1}^h,HP_{q_2}^h) = \frac{N}{h^2}\, w_{q_1}\, w_{q_2}, \tag{42}$$

and it is less than or equal to the sum (29) of class numbers, if certain conditions are satisfied.

The same considerations as in Nos. 5 and 6 with P_0 and P_q exchanged lead to

$$d(P_0,P_q) = \frac{4N}{h^2}\, w_q. \tag{43}$$

This time we need not assume Q_0 to be principal, because Q_q is principal. (43) inserted in (42) and the expressions (23), (29) used, yield further class number relations. But they are proved only under the hypothesis that the P_q are principal. Numerical examples may confirm or refute the assumptions. Unfortunately there are so many restrictions made on d, q_1, q_2 that the simplest possible example would require the knowledge of the class numbers down to $\mathbf{Q}(\sqrt{-18\,480})$.

APPENDIX

CONJECTURES AND PROBLEMS

1. *Conjecture.* Let M be a finite, torsion-free, and reflexive -module with the property that the ranks $L(0,y_0^{n+1}\mathrm{Ext}_h^i(M,h)) = 0$ for $= 1,\cdots$. Then there exists an exponent l such that the submodule M^l f elements of degrees divisible by l with respect to the ring $h^l =$ $_0[y_0^l,\cdots,y_n^l]$ is a free h^l-module.

Maybe that without the last assumption an l exists such that

$$L(\lambda,\mathrm{Ext}^i_{h^l}(M^l,h^l)) = \begin{cases} L(\lambda,\mathrm{Ext}_h^i(M,h)) & \text{for } \lambda = -n-1 \\ 0 & \text{otherwise.} \end{cases}$$

2. *Problem.* From the Theorem of Duality follows: For a reflex-ve and quasifree M there exists a bilinear form $(\varepsilon,\varepsilon^*)$ for

$$\varepsilon \in \mathrm{Ext}_h^i(M,h), \qquad \varepsilon^* \in \mathrm{Ext}_h^{n-i}(M_*,h)$$

ith values in k_0 such that

$$(\varepsilon,\mathrm{Ext}_h^{n-i}(M_*,h)) = 0 \Longleftrightarrow \varepsilon = 0$$

nd

$$(\mathrm{Ext}^i(M,h),\varepsilon^*) = 0 \Longleftrightarrow \varepsilon^* = 0.$$

ive a natural definition for such a bilinear form.

3. *Problem.* Prove an analogue to the reduction lemma of §7 where $= k_0[y_0,\cdots,y_n]$ is replaced by $i_0[y_0,\cdots,y_n]$ with an integral do-ain $i_0 \subset k_0$. The lemma will certainly not hold literally in this ase, too.

Similar considerations are useful in connection with the reduction of a projective variety mod a prime divisor of its constant field.

4. *Problem.* Determine the divisors in which the ring $J(L)$ of modular forms of level L is ramified over the ring $J = J(1)$.

5. *Problem.* In the case of Hilbert modular forms there exist forms obeying the transformation law

$$f(M(z)) \prod_{\nu=1}^{n} \left(\gamma^{\nu} z^{\nu} + \delta^{\nu}\right)^{-h_{\nu}} = f(z)$$

with different exponents h_{ν}. They define in an obvious way reflexive J-ideals and therefore divisors. Use these divisors to obtain knowledge on the structure of J and the field of modular functions and determine Shimizu's rank formula §23, (33). Of course the constants V and $\gamma(w)$ cannot be found explicitly, they will be described as algebraic structure constants of J.

6. *Conjecture.* H. Klingen proved that the image of J under the operator Φ (see §21) is quasireflexive and quasiequal the full ring of modular forms in Z_1. (Zum Darstellungssatz für Siegelsche Modulformen Math. Ztschr. **102**(1967), 30-42.) This image may be even reflexive. A proof must be based on a method of construction of all modular forms of a certain kind. Because of convergence difficulties Eisenstein and Poincaré series cannot be used for small weights. But "generalized" theta series as applied by Eichler (Zur Begründung der Theorie der automorphen Funktionen in mehreren Variablen, Aequationes Math. **3**(1969), 93-111) may serve the purpose.

Many kinds of automorphic forms are obtained by various specializations of Siegel modular forms. In all these cases the images of J may be reflexive.

7. *Conjecture.* Consider Hilbert modular forms in 2 variables with the general transformation law under problem 9. Form the

ndefinite integral

$$F_1(z_1,z_2) = \int_{z_{10}}^{z_1} f(\zeta_1,z_2)\left(\zeta_1-z_1\right)^{h_1-2} d\zeta_1 .$$

t satisfies the functional equations

$$_1(M_1(z_1),M_2(z_2))\left(\gamma_1z_1+\delta_1\right)^{h_1-2}\left(\gamma_2z_2+\delta_2\right)^{-h_2} =$$

$$F_1(z_1,z_2) + P_M(z_1,z_2),$$

here P_M is a polynomial of degree h_1-2 with respect to the first ariable, and

$$_{MN}(z_1,z_2) = P_M(N_1(z_1),N_2(z_2))\left(\gamma_1z_1+\delta_1\right)^{h_1-2}\left(\gamma_2z_2+\delta_2\right)^{-h_2}+P_N(z_1,z_2).$$

urthermore form the integral

$$P_M(z_1,z_2) = \int_{z_{20}}^{z_2} P_M(z_1,\zeta_2)\left(\zeta_2-z_2\right)^{h_2-2} d\zeta_2.$$

hen

$$_{M,N}(z_1,z_2) = P_{MN}(z_1,z_2)$$

$$- P_M(N_1(z_1),N_2(z_2))(\gamma_{N1}z_1+\delta_{N1})^{h_1-2}(\gamma_{N2}z_2+\delta_{N2})^{h_2-2}-P_N(z_1,z_2).$$

s a polynomial in both variables and satisfies the functional equaions for $M,N,R \in \Gamma$:

$$_{M,N}(R_1(z_1),R_2(z_2))\left(\gamma_{R1}z_1+\delta_{R1}\right)^{h_1-2}\left(\gamma_{R2}z_2+\delta_{R2}\right)^{h_2-2} -P_{M,NR} + P_{MN,R}$$

$$- P_{N,R} = 0,$$

xpressing the fact that $P_{M,N}(z)$ is a 2-cocycle in $H^2(\Gamma$, polynomials).

Show that the cohomology groups $H^i(\Gamma$, polynomials) = 0 for $i \neq 2$, nd that the cohomology classes belonging to a modular $f \neq 0$ are $\neq 0$.

Extend the problem to other automorphic forms, also in $n > 2$ ariables.

8. *Problem.* Are the rings of modular forms quasifree modules

with respect to any system of admissible coordinates?

If such a ring J is a quasifree module, is then also every reflexive J-ideal quasifree?

In §16 we made the hypothesis that these rings are even free modules for suitable admissible coordinate rings. The proof that the J are quasifree may be much easier.

9. *A numerical observation.*

For $d = 3$, we have computed the sum of class numbers in (23) for all *admissible* $q < 500$ (Proposition 4) *for which* Γ_q *is the full group of units of a maximal order of the quaternion algebra* Φ_q. These are

$$q = q_1 q_2 = 5.7,\ 5.19,\ 5.31,\ 5.67,\ 5.79,\ 7.17,\ 7.29,\ 7.41,\ 7.53,\ 17.19.$$

In all these cases we found surprisingly that

$$\sum h(4(du^2-q)f^{-2}) = 5d(P_0,P_q) = \frac{5}{3}(q_1-1)(q_2-1). \tag{44}$$

It seems reasonable to conjecture this identity in general under the conditions indicated above. Comparing it with the classical class number relation

$$\sum h((u^2-4q)f^{-2}) = 2q_1(q_2+1) \qquad (q_1 > q_2)$$

it would mean that the class numbers appearing in (44) are in average $\frac{10}{\sqrt{3}}\frac{(q_1-1)(q_2-1)}{q_1(q_2+1)}$ times as large as the main average.

Vol. 74: A. Fröhlich, Formal Groups. IV, 140 pages. 1968. DM 12,–

Vol. 75: G. Lumer, Algèbres de fonctions et espaces de Hardy. VI, 80 pages. 1968. DM 8,–

Vol. 76: R. G. Swan, Algebraic K-Theory. IV, 262 pages. 1968. DM 18,–

Vol. 77: P.-A. Meyer, Processus de Markov: la frontière de Martin. IV, 123 pages. 1968. DM 10,–

Vol. 78: H. Herrlich, Topologische Reflexionen und Coreflexionen. XVI, 166 Seiten. 1968. DM 12,–

Vol. 79: A. Grothendieck, Catégories Cofibrées Additives et Complexe Cotangent Relatif. IV, 167 pages. 1968. DM 12,–

Vol. 80: Seminar on Triples and Categorical Homology Theory. Edited by B. Eckmann. IV, 398 pages. 1969. DM 20,–

Vol. 81: J.-P. Eckmann et M. Guenin, Méthodes Algébriques en Mécanique Statistique. VI, 131 pages. 1969. DM 12,–

Vol. 82: J. Wloka, Grundräume und verallgemeinerte Funktionen. VIII, 131 Seiten. 1969. DM 12,–

Vol. 83: O. Zariski, An Introduction to the Theory of Algebraic Surfaces. IV, 100 pages. 1969. DM 8,–

Vol. 84: H. Lüneburg, Transitive Erweiterungen endlicher Permutationsgruppen. IV, 119 Seiten. 1969. DM 10,–

Vol. 85: P. Cartier et D. Foata, Problèmes combinatoires de commutation et réarrangements. IV, 88 pages. 1969. DM 8,–

Vol. 86: Category Theory, Homology Theory and their Applications I. Edited by P. Hilton. VI, 216 pages. 1969. DM 16,–

Vol. 87: M. Tierney, Categorical Constructions in Stable Homotopy Theory. IV, 65 pages. 1969. DM 6,–

Vol. 88: Séminaire de Probabilités III. IV, 229 pages. 1969. DM 18,–

Vol. 89: Probability and Information Theory. Edited by M. Behara, K. Krickeberg and J. Wolfowitz. IV, 256 pages. 1969. DM 18,–

Vol. 90: N. P. Bhatia and O. Hajek, Local Semi-Dynamical Systems. II, 157 pages. 1969. DM 14,–

Vol. 91: N. N. Janenko, Die Zwischenschrittmethode zur Lösung mehrdimensionaler Probleme der mathematischen Physik. VIII, 194 Seiten. 1969. DM 16,80

Vol. 92: Category Theory, Homology Theory and their Applications II. Edited by P. Hilton. V, 308 pages. 1969. DM 20,–

Vol. 93: K. R. Parthasarathy, Multipliers on Locally Compact Groups. III, 54 pages. 1969. DM 5,60

Vol. 94: M. Machover and J. Hirschfeld, Lectures on Non-Standard Analysis. VI, 79 pages. 1969. DM 6,–

Vol. 95: A. S. Troelstra, Principles of Intuitionism. II, 111 pages. 1969. DM 10,–

Vol. 96: H.-B. Brinkmann und D. Puppe, Abelsche und exakte Kategorien, Korrespondenzen. V, 141 Seiten. 1969. DM 10,–

Vol. 97: S. O. Chase and M. E. Sweedler, Hopf Algebras and Galois theory. II, 133 pages. 1969. DM 10,–

Vol. 98: M. Heins, Hardy Classes on Riemann Surfaces. III, 106 pages. 1969. DM 10,–

Vol. 99: Category Theory, Homology Theory and their Applications III. Edited by P. Hilton. IV, 489 pages. 1969. DM 24,–

Vol. 100: M. Artin and B. Mazur, Etale Homotopy. II, 196 Seiten. 1969. DM 12,–

Vol. 101: G. P. Szegö et G. Treccani, Semigruppi di Trasformazioni Multivoche. VI, 177 pages. 1969. DM 14,–

Vol. 102: F. Stummel, Rand- und Eigenwertaufgaben in Sobolewschen Räumen. VIII, 386 Seiten. 1969. DM 20,–

Vol. 103: Lectures in Modern Analysis and Applications I. Edited by C. T. Taam. VII, 162 pages. 1969. DM 12,–

Vol. 104: G. H. Pimbley, Jr., Eigenfunction Branches of Nonlinear Operators and their Bifurcations. II, 128 pages. 1969. DM 10,–

Vol. 105: R. Larsen, The Multiplier Problem. VII, 284 pages. 1969. DM 18,–

Vol. 106: Reports of the Midwest Category Seminar III. Edited by S. Mac Lane. III, 247 pages. 1969. DM 16,–

Vol. 107: A. Peyerimhoff, Lectures on Summability. III, 111 pages. 1969. DM 8,–

Vol. 108: Algebraic K-Theory and its Geometric Applications. Edited by R. M. F. Moss and C. B. Thomas. IV, 86 pages. 1969. DM 6,–

Vol. 109: Conference on the Numerical Solution of Differential Equations. Edited by J. Ll. Morris. VI, 275 pages. 1969. DM 18,–

Vol. 110: The Many Facets of Graph Theory. Edited by G. Chartrand and S. F. Kapoor. VIII, 290 pages. 1969. DM 18,–

Vol. 111: K. H. Mayer, Relationen zwischen charakteristischen Zahlen. III, 99 Seiten. 1969. DM 8,–

Vol. 112: Colloquium on Methods of Optimization. Edited by N. N. Moiseev. IV, 293 pages. 1970. DM 18,–

Vol. 113: R. Wille, Kongruenzklassengeometrien. III, 99 Seiten. 19 DM 8,–

Vol. 114: H. Jacquet and R. P. Langlands, Automorphic Forms on GL VII, 548 pages. 1970. DM 24,–

Vol. 115: K. H. Roggenkamp and V. Huber-Dyson, Lattices over Orde XIX, 290 pages. 1970. DM 18,–

Vol. 116: Séminaire Pierre Lelong (Analyse) Année 1969. IV, 195 pa 1970. DM 14,–

Vol. 117: Y. Meyer, Nombres de Pisot, Nombres de Salem et Ana Harmonique. 63 pages. 1970. DM 6,–

Vol. 118: Proceedings of the 15th Scandinavian Congress, Oslo 19 Edited by K. E. Aubert and W. Ljunggren. IV, 162 pages. 1970. DM 12,–

Vol. 119: M. Raynaud, Faisceaux amples sur les schémas en grou et les espaces homogènes. III, 219 pages. 1970. DM 14,–

Vol. 120: D. Siefkes, Büchi's Monadic Second Order Succes Arithmetic. XII, 130 Seiten. 1970. DM 12,–

Vol. 121: H. S. Bear, Lectures on Gleason Parts. III, 47 pages. 1 DM 6,–

Vol. 122: H. Zieschang, E. Vogt und H.-D. Coldewey, Flächen ebene diskontinuierliche Gruppen. VIII, 203 Seiten. 1970. DM 16

Vol. 123: A. V. Jategaonkar, Left Principal Ideal Rings. VI, 145 pa 1970. DM 12,–

Vol. 124: Séminare de Probabilités IV. Edited by P. A. Meyer. IV, pages. 1970. DM 20,–

Vol. 125: Symposium on Automatic Demonstration. V, 310 pages. 1 DM 20,–

Vol. 126: P. Schapira, Théorie des Hyperfonctions. XI, 157 pages. 1 DM 14,–

Vol. 127: I. Stewart, Lie Algebras. IV, 97 pages. 1970. DM 10,–

Vol. 128: M. Takesaki, Tomita's Theory of Modular Hilbert Alge and its Applications. II, 123 pages. 1970. DM 10,–

Vol. 129: K. H. Hofmann, The Duality of Compact Semigroups C*- Bigebras. XII, 142 pages. 1970. DM 14,–

Vol. 130: F. Lorenz, Quadratische Formen über Körpern. II, 77 Se 1970. DM 8,–

Vol. 131: A Borel et al., Seminar on Algebraic Groups and Rel Finite Groups. VII, 321 pages. 1970. DM 22,–

Vol. 132: Symposium on Optimization. III, 348 pages. 1970. DM 22

Vol. 133: F. Topsøe, Topology and Measure. XIV, 79 pages. 19 DM 8,–

Vol. 134: L. Smith, Lectures on the Eilenberg-Moore Spectral Seque VII, 142 pages. 1970. DM 14,–

Vol. 135: W. Stoll, Value Distribution of Holomorphic Maps into C pact Complex Manifolds. II, 267 pages. 1970. DM 18,–

Vol. 136: M. Karoubi et al., Séminaire Heidelberg-Saarbrücken-S buorg sur la K-Théorie. IV, 264 pages. 1970. DM 18,–

Vol. 137: Reports of the Midwest Category Seminar IV. Edite S. MacLane. III, 139 pages. 1970. DM 12,–

Vol. 138: D. Foata et M. Schützenberger, Théorie Géométrique Polynômes Eulériens. V, 94 pages. 1970. DM 10,–

Vol. 139: A. Badrikian, Séminaire sur les Fonctions Aléatoires L aires et les Mesures Cylindriques. VII, 221 pages. 1970. DM 18

Vol. 140: Lectures in Modern Analysis and Applications II. Edite C. T. Taam. VI, 119 pages. 1970. DM 10,–

Vol. 141: G. Jameson, Ordered Linear Spaces. XV, 194 pages. 1 DM 16,–

Vol. 142: K. W. Roggenkamp, Lattices over Orders II. V, 388 pa 1970. DM 22,–

Vol. 143: K. W. Gruenberg, Cohomological Topics in Group The XIV, 275 pages. 1970. DM 20,–

Vol. 144: Seminar on Differential Equations and Dynamical Syst II. Edited by J. A. Yorke. VIII, 268 pages. 1970. DM 20,–

Vol. 145: E. J. Dubuc, Kan Extensions in Enriched Category Th XVI, 173 pages. 1970. DM 16,–

Vol. 146: A. B. Altman and S. Kleiman, Introduction to Grothend Duality Theory. II, 192 pages. 1970. DM 18,–

Vol. 147: D. E. Dobbs, Cech Cohomological Dimensions for mutative Rings. VI, 176 pages. 1970. DM 16,–

Vol. 148: R. Azencott, Espaces de Poisson des Groupes Localer Compacts. IX, 141 pages. 1970. DM 14,–

Vol. 149: R. G. Swan and E. G. Evans, K-Theory of Finite Groups Orders. IV, 237 pages. 1970. DM 20,–